Aging - Life Span and Life Expectancy

Edited by Robert J. Reynolds and Steven M. Day

Published in London, United Kingdom

IntechOpen

Supporting open minds since 2005

Aging - Life Span and Life Expectancy
http://dx.doi.org/10.5772/intechopen.77734
Edited by Robert J. Reynolds and Steven M. Day

Contributors
Prisilla Jayanthi, Iyyanki Murali Krishna, Aborlo Kennedy Nkporbu, Alvaro Soto, Marcelo Peldoza, Debora Pollak, Michael John Stones, Bahram H Arjmandi, Elizabeth Marie Foley, Sarah Worobetz, Jason Randle, Carlina Marchese, Shauna Fossum, Dane Ostrom, Peter Brink

First published in London, United Kingdom, 2020 by IntechOpen
IntechOpen is the global imprint of INTECHOPEN LIMITED, registered in England and Wales, registration number: 11086078, 7th floor, 10 Lower Thames Street, London, EC3R 6AF, United Kingdom
Printed in Croatia

British Library Cataloguing-in-Publication Data
A catalogue record for this book is available from the British Library

Additional hard and PDF copies can be obtained from orders@intechopen.com

Aging - Life Span and Life Expectancy
Edited by Robert J. Reynolds and Steven M. Day
p. cm.
Print ISBN 978-1-78984-081-0
Online ISBN 978-1-78984-082-7
eBook (PDF) ISBN 978-1-83880-915-7

For EU product safety concerns:
IN TECH d.o.o., Prolaz Marije Krucifikse Kozulić 3, 51000 Rijeka, Croatia, info@intechopen.com or visit our website at intechopen.com.

Meet the editors

Dr. Reynolds's research has focused on mortality and life expectancy in occupational groups (such as professional athletes, astronauts, and fashion models), and of persons with a variety of medical conditions affecting life expectancy, including mental illness, cerebral palsy, and other developmental disabilities. He has authored numerous peer-reviewed publications and has presented papers or posters at professional meetings of academic organizations, including the American Public Health Association (APHA), the Texas Public Health Association, the Society for Epidemiologic Research (SER), the Society for Pediatric and Perinatal Research (SPER), and the American Academy for Cerebral Palsy and Developmental Medicine (AACPDM). Dr. Reynolds is a member of the American Statistical Association, SER, the Aerospace Medical Association, and a Fellow of the AACPDM.

Dr. Day earned his PhD in applied statistics with substantive field epidemiology from the University of California at Riverside in 2001. He is the President and CEO of Mortality Research & Consulting, Inc. The company provides expert consulting services related to statistics and epidemiology, primarily focusing on mortality, survival, and life expectancy. Dr. Day has co-authored numerous peer-reviewed articles and book chapters related to survival, mortality, life expectancy, and other epidemiological topics. His research has focused on specific occupational cohorts (e.g., professional athletes and astronauts), and on people with mental health issues, developmental disabilities, traumatic brain injuries, spinal cord injuries, epilepsy, and other medical conditions. He is a member of the Society for Epidemiologic Research, the American Academy for Cerebral Palsy and Developmental Medicine, the American Statistical Association, and the Royal Society of Medicine. He has presented papers, posters, and instructional courses on the topic of life expectancy and related issues at professional meetings in the US, Canada, and the UK.

Contents

Preface **XI**

Chapter 1 **1**
Psychotropic Medication Use and Mortality in Long-Term Care Residents
by Michael J. Stones, Sarah Worobetz, Jason Randle, Carlina Marchese, Shauna Fossum, Dane Ostrom and Peter Brink

Chapter 2 **25**
Evidence for the Effectiveness of Soy in Aging and Improving Quality of Life
by Bahram Herman Arjmandi and Elizabeth Marie Foley

Chapter 3 **49**
Epidemiology and Management of Intracerebral Hemorrhage in Chile
by Álvaro Soto, Marcelo Peldoza and Debora Pollak

Chapter 4 **61**
Impact of Chronic Medical and Neuropsychiatric Illnesses on Quality of Life and Life Expectancy among Patients at the University of Port Harcourt Teaching Hospital (UPTH)
by Aborlo Kennedy Nkporbu

Chapter 5 **83**
Crude Birth Rate and Crude Mortality Rate in India: A Case of Application of Regression in Healthcare
by Prisilla Jayanthi and Muralikrishna Iyyanki

Preface

This edited volume is a collection of reviewed and relevant research chapters concerning the developments within the "Aging - Life Span and Life Expectancy" field of study. The book includes scholarly contributions by various authors and has been edited by an expert in the field. Each contribution comes as a separate chapter complete in itself but directly related to the book's topics and objectives.

The book includes chapters dealing with the following topics: Psychotropic Medication Use and Mortality in Long-Term Care Residents; Evidence for the Effectiveness of Soy in Aging and Improving Quality of Life; Epidemiology and Management of Intracerebral Hemorrhage in Chile; Impact of Chronic Medical and Neuropsychiatric Illnesses on Quality of Life and Life Expectancy among Patients at the University of Port Harcourt Teaching Hospital (UPTH); Crude Birth Rate and Crude Mortality Rate in India: A Case of Application of Regression in Healthcare.

The target audience comprises scholars and specialists in the field.

InTechOpen

Chapter 1

Psychotropic Medication Use and Mortality in Long-Term Care Residents

Michael J. Stones, Sarah Worobetz, Jason Randle, Carlina Marchese, Shauna Fossum, Dane Ostrom and Peter Brink

Abstract

This chapter examines associations between psychotropic medications and mortality in long-term care home (LTCH) settings. We report new findings with census-level data from all new admissions to long-term care homes in the province of Ontario, Canada (i.e., 20,414 new residents). The data include three linked sets that indicate mortality during the financial years 2010–2011 and 2011–2012. One dataset, the Resident Assessment Instrument 2.0 (RAI 2.0), provides information on demographics, functional capability, clinical conditions, clinical diagnoses, mortality risk, and psychotropic medications. The latter include antipsychotics, antidepressants, analgesics, anxiolytics, and hypnotics. Administration of the RAI 2.0 occurs at resident intake, at quarterly intervals and annually. New analyses reported here examine predictors of daily and *pro re nata* (i.e., PRN or "as needed") prescriptions of psychotropic medications. However, the most important analyses concern predictors of mortality within intervals of up to 90 days from the final RAI 2.0 assessment. After control for confounding variables, the findings indicate (1) attenuated mortality with daily prescription of frequently prescribed psychotropics (i.e., antipsychotics, antidepressants, and analgesics), (2) augmented mortality with PRN prescriptions for each type of psychotropic medication, and (3) evidence that PRN prescribing overturns beneficial effects of daily prescriptions, whereas the latter reduces the deleterious effects of PRN prescribing.

Keywords: mortality, medication, psychotropic, antipsychotic, analgesic, antidepressant, anxiolytic, hypnotic, aging, elderly, gerontology, long-term care, dementia

1. Introduction

This chapter illustrates a truism that pathways pursued in scientific investigation often deviate from linear trajectories. The research we report here evolved from concerns about ongoing practices in the continuing care of older people, a serendipitous convergence of people with compatible research needs and desires, and totally unanticipated findings in need of further investigation. This section of the chapter traces the earlier stages of that progression.

1.1 Concern about associations of antipsychotic medication and mortality in older people

The research that follows evolved from participation by the lead author in unpublished work in the late 1990s commissioned by the Canadian Institute for Health Information (CIHI). The purpose was to analyze early Canadian data on version two of the Minimum Data Set (MDS 2.0), which is the former name of the RAI 2.0. In 1996, this tool became mandated for use in all chronic care hospitals, now known as complex community care (CCC) facilities in the Canadian Province of Ontario. The residents of these facilities are mainly older people in receipt of continuing care and/or rehabilitation associated with disabling chronic illness.

The findings from that work that was most troubling included high frequencies of physical restraint and chemical management in Canadian facilities compared to findings with the same tool in other countries. Although Canadian physical restraint levels lessened in frequency since that time, such is not the case for chemical management. Hence the enduring interest in chemical management by these authors.

The purpose of chemical management is to address symptoms that fall under the umbrella of behavioral and psychological symptoms of dementia (BPSD). The definition of the latter at a 1996 Consensus Conference of the International Psychogeriatric Association (IPA) is as follows: "symptoms of disturbed perception, thought content, mood or behavior that frequently occur in patients with dementia" [1]. The behavioral symptoms include physical aggression, loud vocalization, restlessness, agitation, and wandering. The psychological symptoms include anxiety, depressive mood, hallucinations, and delusions.

Current estimates suggest that over half the patients with dementia are at risk to such symptoms, which typically arise during the middle or later stages of the disease. Chemical management may include analgesics to lessen pain and discomfort, along with antidepressants, anti-anxiety medication and antipsychotics, all of which address behavioral and psychological symptoms [2]. The most frequent concerns about chemical management relate to antipsychotic medication. These are drugs first developed for the treatment of psychosis [3]. Presently, antipsychotic drugs fall within two categories. Typical antipsychotics include those initially developed to treat psychosis, while atypical antipsychotics were developed later to reduce adverse side effects of the former.

Concerns about harmful effects of antipsychotics in dementia patients are legitimate. The adverse effects of these drugs include high rates of cardiovascular events, cardiac arrhythmias, cerebrovascular events, cognitive decline, extrapyramidal symptoms, pneumonia, falls and fractures, and others [4]. However, the most serious concern is an elevated risk of mortality. Notice of such concerns began early this millennium with evidence that these adverse effects were over and above those associated with old age, an underlying dementia, and behavioral and psychological symptoms that might precipitate the use of antipsychotics [5].

In 2002, the manufacturer of a typical antipsychotic medication warned of an increased risk of adverse cardiovascular events [5]. Subsequently, the US Federal Food and Drug Administration (FDA) required "black box" warnings about the use of atypical antipsychotics (in 2005) and typical antipsychotics (in 2008). The warning states: "WARNING: Elderly patients with dementia-related psychosis treated with antipsychotic drugs are at an increased risk of death." Health authorities in other countries subsequently expressed similar concerns.

Conclusions from the most recent and most extensive meta-analysis of studies that relate the use antipsychotic medication to the mortality of elderly people [6] are as follows. Mortality risk with antipsychotic medication (1) is twice that of people without such prescription, (2) comparable between typical and atypical antipsychotics, (3) highest during the initial half-year of use, (4) higher at higher dosage, and (5) comparable between people with or without diagnosed dementia. One interpretation of these five points is that people with more severe dementia-related psychosis are at greater risk of mortality, with that risk lessening 6 months after they begin that course of medication. However, findings from placebo-controlled trials indicate antipsychotic use is associated with an increase in mortality above and beyond the baseline dementia symptoms. The authors of this meta-analysis also recommended a restricted use of antipsychotic medication with older people and encouraged the deployment of de-prescribing practices.

Final thoughts by those authors concur with comments made earlier by authors of this chapter [7]. Ralph and Espinet [6] anticipate greater cultural disapproval about the sedation of older people through antipsychotic, anxiolytic, and hypnotic medications. They envision attitudinal changes within the health and legal professions to consider such practices examples of systemic elder abuse that requires legal reform. A current drive toward the de-prescribing of antipsychotic medication to elderly people is consistent with these beliefs [8].

1.2 Preliminary study of antipsychotic medication and mortality in older people

When Sarah Worobetz, a doctoral student at Lakehead University, wanted to research relationships between antipsychotic medication and mortality in older people, after control for a wide range of variables we describe subsequently, faculty members Michael Stones and Peter Brink were happy to oblige. These researchers had a working familiarity with the RAI 2.0. They hoped to obtain census level data, with linkages to other mortality relevant datasets, to provide Sarah with the means to conduct her research. Peter Brink successfully submitted a proposal to CIHI for access to access these data.

The RAI 2.0 is a standardized assessment tool used routinely in LTCH and CCC facilities in Ontario and other settings across the world. The tool contains over 350 items relevant to medical diagnoses; physical, cognitive, social, and emotional functioning; and treatment categories that include medication use. It also indexes mortality within the relevant facilities. The trained health care professionals responsible for RAI 2.0 assessments obtain that information from multiple sources, such as direct observation, medical records, and communication with family members and other health care professionals. Objective scales on the RAI 2.0 consist of sets of items selected for relevance to a given construct. Evaluation of such scales may be against "gold standard" measures of the constructs (e.g., measures of activities of daily living, cognitive status, depression, aggression, and pain) or relevant outcomes (e.g., mortality risk). From a measurement perspective, previous findings on data quality and the reliability and validity of RAI 2.0 measures are positive [8–10].

The antipsychotic medication item on the RAI 2.0 falls within a psychotropic category that also contains items on antidepressant, analgesic, anxiolytic, and hypnotic medications. The wording of each of these items asks for the number of days during the past week that the resident received the medication. This form of measurement differs from that common to previous studies of psychotropic medication and mortality, which invariably report specific medications and dosages but not frequency of medication use.

The other databases provided by CIHI are the Discharge Abstract Database (DAD) and the National Ambulatory Care Reporting System (NACRS). The DAD contains demographic, administrative, and clinical data for hospital discharges (inpatient acute, chronic, and rehabilitation) and day surgeries. The NACRS contains data for hospital-based and community-based emergency and ambulatory care (e.g., day surgery and outpatient clinics). Both datasets contain mortality data pertaining to their respective contexts. CIHI encrypted the personal and facility identifiers across datasets to ensure anonymity of residents. Brink merged these three datasets for purposes of analysis. Consequently, the merged file contains mortality data both within LTCH and CCC facilities and for those discharged to other health care settings.

The dataset analyzed by Worobetz includes all admission, quarterly and annual RAI 2.0 assessment for residents aged over 65 years in all LTCH and CCC facilities in Ontario during the financial years 2010–2011 and 2011–2012. The data are from 102,658 residents of approximately 760 facilities with a mean of 5.83 assessments per resident. Approximately 70% of residents are female and 30% male, with a mean age over all assessments of approximately 84 years. Approximately 86% of residents live in LTCH and 14% in CCC facilities. The mean length of stay prior to the first assessment was approximately 20 months, with the age at entry approximately 82 years. The distribution of antipsychotic prescriptions shows that approximately 69% of residents are without medication, 29% have daily prescriptions, and 2% have prescriptions of 1–6 days per week, which for purposes of this research we describe as PRN prescriptions. The total mortality rate during the 2-year period of data collection is approximately 32%.

The initial findings by Worobetz on relationships between mortality and type of antipsychotic prescription came as a big surprise. Her analysis by generalized linear mixed modeling (GLMM) appointed LTCH as a random effect variable (i.e., independent entities) with residents clustered (i.e., showing covariation) within their respective facilities. Such modeling is appropriate because of differences among facilities with respect to admission criteria, population size, staffing levels, types of programming, treatment protocols, etc., with localized interpersonal exchanges within facilities that foster covariation among residents. Compared to residents with no prescription for antipsychotic medication, her findings show attenuated mortality for those with daily prescription, but augmented mortality for those with PRN prescription. A possible interpretation is that these findings are consistent with earlier evidence of a protective effect of antipsychotics after 6 months but an increased mortality risk for residents prescribed antipsychotics on a PRN basis because they began to exhibit relevant symptoms.

Subsequent GLMM analyses by Worobetz included all residents, only those from LTCH, only those from CCC, new admissions, residents with dementia, and combinations of the preceding. The fixed effect variables in such analyses included not just prescriptions for antipsychotic medication but multivariate control for confounding variables such as demographics, scores on RAI 2.0 scales (e.g., activity limitation, cognitive status, aggression, depression, and mortality risk), temporal changes on those scales, and medical diagnoses (e.g., cancer, dementia, maniac depression, and schizophrenia [11]). The findings from all these analyses consistently show highest mortality among residents with PRN prescriptions on the final assessment.

1.3 Studies of other psychotropic medications and mortality in older people

The preceding findings provide reasons to broaden the scope of investigation to encompass mortality in relation to prescriptions of other types of psychotropic medication. We begin this section with brief discussion of prescribing practices and mortality associated with analgesics, antidepressants, anxiolytics and hypnotics, which in the RAI 2.0 fall within an item-set of psychotropic medications. Then follows discussion of problems associated with PRN prescribing practices. Finally, we report findings from separate analyses of mortality against these types of psychotropic medications.

Prescribing practices with analgesics show the following trends. Although rates for PRN prescription in elderly care services are generally low, some reports indicate highest levels for analgesics [12]. Worldwide, scheduled rates for analgesic use (that include acetaminophen and opioids) in LTCH show a historical increase, whereas rates for scheduled plus PRN rates show no such increase [13]. Recent findings from the Czech Republic suggest that a large proportion of LTCH residents with pain receive no analgesic medication and a moderate proportion of those that receive analgesic medication continue to report pain. These findings of analgesic under-prescription are consistent with those from North America and elsewhere in Europe. The lowest frequency of reported pain and lowest prevalence of analgesic administration are for residents with moderate-to-severe dementia [14], which suggests this group's susceptibility to under-detection and under-prescribing of this medication.

Anti-depressant medications find frequent use in older people, with average prevalence rates of approximately 25% [15, 16]. Recent evidence suggests no association between antidepressant prescriptions and augmented risk of all-cause mortality [17, 18]. However, best practice guidelines recommend caution when prescribing because low adherence may increase risks of fatal cardiovascular and cerebrovascular injuries [19]. On the other hand, high adherence appears to lower mortality risk [20]. The findings give rise to hypotheses that intermittent use of antidepressants (e.g., comparable to PRN prescribing) has unfavorable implications for mortality whereas regular use (e.g., associated with daily prescription) has favorable implications. However, we will discuss other interpretations later in the chapter.

A recent review suggests that benzodiazepines are the most frequently prescribed anxiolytic medications for geriatric anxiety [21]. However, consensus is low about whether anxiolytic and hypnotic medications have unfavorable implications for mortality risk amongst older adults [22]. On the other hand, a large-scale retrospective cohort study of patients in UK primary care concluded that anxiolytic and hypnotic drugs were associated with significantly increased risk of mortality over a 7-year period, after adjusting for a range of potential confounders [23].

A number of studies and reviews examined PRN prescription in psychiatric and LTCH settings [24–27]. Summary findings indicate higher PRN use for residents with lower care needs, frequent use alongside regularly scheduled medications, and recent entry into a facility. Contextual factors also have a strong influence on PRN prescribing. These include general levels of activity and disturbance on the ward, staffing level, perceived competence of staff, and familiarity of the staff with residents. The reports also indicate frequent omissions and errors in records of PRN usage. Findings that relate mortality to PRN usage appear to be absent in the literature.

Following the thesis research by Worobetz, subsequent analyses of the same dataset examined relationships of mortality to prescriptions of anesthetics (Jason Randle), antidepressants (Carlina Marchese and Shauna Fossum), anxiolytics (Michael Stones), and hypnotics (Dane Ostrom). We describe the sampling procedure in the following section. The main findings show significantly higher mortality with PRN prescriptions for each type of psychotropic medication compared with daily or null prescription. These findings persist even after statistical control of relevant confounding variables. They provide the impetus for the new analyses that follow.

2. New analyses on relationships between psychotropic medication and mortality

We try here to expand the scope and level of precision beyond those present in previous analyses, each of which examined associations of mortality with a single psychotropic medication. First, we analyze the effects on mortality of all the psychotropic prescriptions within a series of multivariate analyses that includes the psychotropic medications. Second, we introduce a verified measure of mortality risk into the array of control variables. Third, we examine intervals for mortality of 90 days from the final RAI 2.0 assessment (i.e., the scheduled date of the next assessment) and shorter periods within that interval. Fourth, we examine predictors of daily and PRN prescriptions of psychotropic medications. Fifth, we examine the effects on mortality for residents receiving only daily, only PRN, or both daily and PRN prescriptions of psychotropic medication.

2.1 Methods

Section 1.2 described the three datasets linked and encrypted by CIHI. Here we report analyses that restrict data entry to (1) scheduled intake, quarterly and yearly assessments, (2) in LTCH settings, (3) for residents aged 65+ years, (4) with an intake assessment during the financial year 2010/2011, and (5) subsequent assessments that do not exceed first yearly assessment (i.e., <13 months after the intake assessment). Consequently, the data enable computation of a census level 1-year incidence rate for mortality among LTCH residents aged 65 years and older.

The analyses that follow begin at a descriptive level and proceed to an inferential level. The latter analyses were performed using the SPSS 25 GLMM program. The target variables in different analyses include mortality within 90 days of the final assessment, time to mortality within that interval, and frequencies of daily and PRN prescriptions for psychotropic medication categories. The random variable in all analyses identifies the LTCH at intake assessment. Evidence of statistical significance for the random variable would confirm that levels on the target variable differ across the LTCH spectrum.

The fixed effects in different analyses include the following RAI 2.0 measures: demographics (gender, age at assessment); scales (Activities of Daily Living Hierarchy, Cognitive Performance Scale, Depression Rating Scale, Aggressive Behavior Scale, Pain Scale and the Changes in Health, End-Stage Disease, and Symptoms and Signs Scale); diagnoses (insomnia and dementia), and medications (antipsychotic, antidepressant, analgesic, anxiolytic, and hypnotic). Convincing evidence from earlier and more recent publications [28] attest to good data quality and psychometric properties associated with RAI 2.0 measures.

2.2 Results

Here we describe the sample with respect to demographics, psychotropic medication use, and mortality. The sample of 20,414 residents of 631 LTCH includes approximately 38% admissions from inpatient acute care, 28% from a private home, 13% from residential board and care, and 10% from home care. The remaining 11% of admissions are from residences with 24-hour nursing care, inpatient continuing care, inpatient rehabilitation, or inpatient psychiatry. The sample comprises 33.6% men and 66.4% women. The age distribution for men has a mean of 83.03 years and standard deviation of 7.37 years, with respective estimates for women of 85.29 and 7.19 years.

Table 1 shows percentage frequencies for residents prescribed psychotropic medication during the week before the final assessment. Daily prescription is highest for analgesics followed by antidepressants followed by antipsychotics. More than half the residents use analgesics daily, nearly half use antidepressants daily, and approximately 30% receive daily antipsychotics. The frequency of PRN usage is low. The highest frequencies are 7.5% for analgesics and 3.3% for anxiolytics.

Frequencies of daily prescriptions for different types of psychotropic medication are as follows: 33.4% of residents receive one medication; 31.5% receive two medications; 17.6% use three or four (i.e., 3+) medications; and 17.4% of residents have no daily psychotropic medication. Frequencies for PRN prescriptions of psychotropic medication are as follows: 10.9% of residents use one medication; 1.6% use two or three (i.e., 2+) medications; and 87.5 of residents have no

	Prescription of psychotropic medication		
Medication type	**None (%)**	**PRN (%)**	**Daily (%)**
Antipsychotic	69.2	1.4	29.4
Antidepressant	50.9	1.4	47.7
Analgesic	34.6	7.5	57.9
Anxiolytic	84.9	3.3	11.8
Hypnotic	93.7	0.7	5.6

Table 1.
Percentage frequencies for usage of psychotropic medications.

psychotropic medication on a PRN basis. Frequencies for combinations of daily and PRN prescriptions are as follows: 74.6% of residents have only daily prescriptions for psychotropic medications; 4.5% have only PRN prescriptions; 8.0% have both daily and PRN prescriptions; and 12.9% have no prescription. These findings indicate that about half the residents receive two or more psychotropic medications daily, 12.5% receive one or more on a PRN basis, and 8.0% have both daily and PRN prescriptions.

Probabilistic mortalities within 90 days of the final assessment are 18.1% overall, 21.1% for males and 16.3% for females. The percentages of residents dying within different time periods after the final assessment are as follows: 2.2% mortality within 7 days after the final assessment, 6.1% within 8–30 days, and 9.8% within 31–90 days. The proportion of residents with no mortality within 90 days is 82%.

2.2.1 Analysis of mortality's association with prescriptions of psychotropic medications

The purpose is to evaluate whether findings from separate analyses of relationships between mortality and types of psychotropic medication replicate in multivariate analysis that includes all such medications and potential confounders. The most significant findings from the earlier research indicate highest mortality with PRN prescription for each type of psychotropic medication. Replication in multivariate analysis would suggest that such relationships exist independently of any covariation in prescribing practices across types of medication.

The target variable in this analysis is mortality within 90 days of the final assessment, with 90 days being the time before the next scheduled assessment. The reference category for this variable is absence of mortality. The distribution of the target variable is binomial and related to a linear model by a complementary log-log link. SPSS 25 recommends such a linkage in survival analysis, where some observations have no termination event.

The fixed effects include demographic variables (i.e., men/women, age at the final assessment); measures of functional capability (i.e., Cognitive Performance Scale, Activities of Daily Living Hierarchy); and prescribed frequency of usage for each type of psychotropic medication (i.e., none, PRN, daily). The analysis accords with conventional GLMM practices that include centering of continuous measures on their grand mean and comparison of levels on a nominal variable with a reference category. For the present nominal measures, the respective reference categories are male gender and zero frequency of usage for a psychotropic medication.

Findings for the random variable in this and all subsequent analyses indicate significant differences (at $p < 0.001$) in levels of the target variable of mortality across facilities. The findings for fixed effects in **Table 2** indicate lower mortality in women than men and higher mortality at older ages. Mortality also increases with

Model term	Odds ratio	Sig.	95% confidence interval	
			Lower	Upper
Intercept	0.21	0.000	0.19	0.23
Female	0.68	0.000	0.64	0.73
Male				
Age at assessment	1.03	0.000	1.03	1.04
Activities of daily living	1.49	0.000	1.46	1.53
Cognitive performance	0.98	0.144	0.96	1.01
Daily antipsychotic	0.92	0.040	0.85	1.00
PRN antipsychotic	1.59	0.000	1.28	1.96
No antipsychotic				
Daily antidepressant	0.78	0.000	0.72	0.83
PRN antidepressant	1.37	0.004	1.11	1.69
No antidepressant				
Daily analgesic	0.98	0.555	0.91	1.05
PRN analgesic	1.72	0.000	1.53	1.93
No analgesic				
Daily anxiolytic	0.99	0.899	0.89	1.11
PRN anxiolytic	1.40	0.000	1.20	1.64
No anxiolytic				
Daily hypnotic	1.08	0.301	0.93	1.25
PRN hypnotic	1.77	0.000	1.31	2.40
No hypnotic				

Table 2.
Fixed effect odds ratios including daily, PRN and No psychotropic predictors of mortality within 90 days of the final assessment.

activity limitation, as indexed by higher scores on the activities of daily living scale. The findings for psychotropic medications show significantly lower mortality for daily antipsychotic and antidepressant prescription compared with no prescriptions. The findings for PRN prescriptions show significantly higher mortality for all types of psychotropic medication relative to no prescription and (in subsequent paired comparisons) daily prescription.

Figures 1–5 illustrate the magnitude of these findings, which are present in both men and women.

The findings from this analysis replicate in a multivariate context those from separate analyses of each psychotropic medication. We conclude, therefore, that associations between PRN prescription and higher mortality are independent for each type of psychotropic medication. Compared to previous research, the findings of lower mortality with daily antipsychotic and antidepressant prescription are unusual for the former but not the latter. However, we have no reason to doubt their validity, based as they are on census level data from an entire province. It is possible that confounding factors that contribute to earlier reports of augmented mortality with antipsychotic medication include failures to distinguish daily from intermittent usage and to identify adherence to daily prescription regimens. The presence of either confound might overturn protective effects associated with antipsychotic medication.

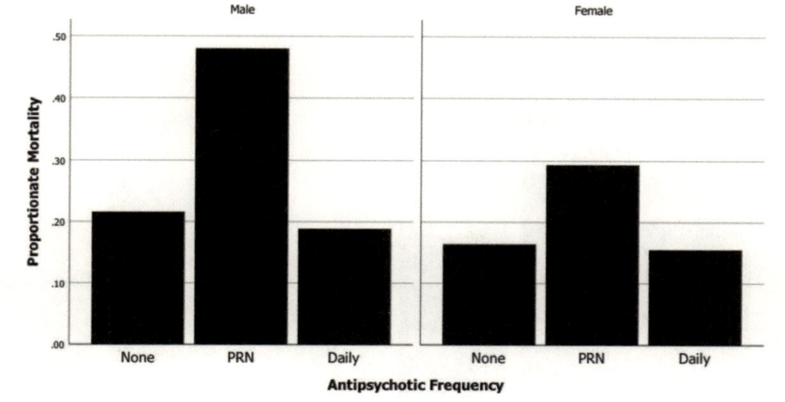

Figure 1.
Proportionate mortality by antipsychotic frequency in men and women.

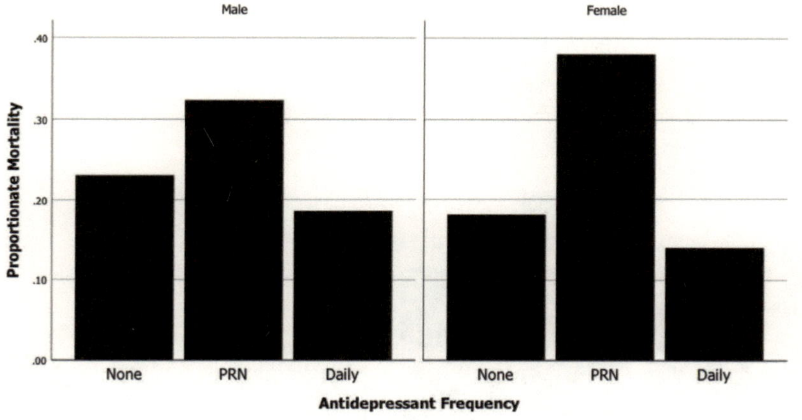

Figure 2.
Proportionate mortality by antidepressant frequency in men and women.

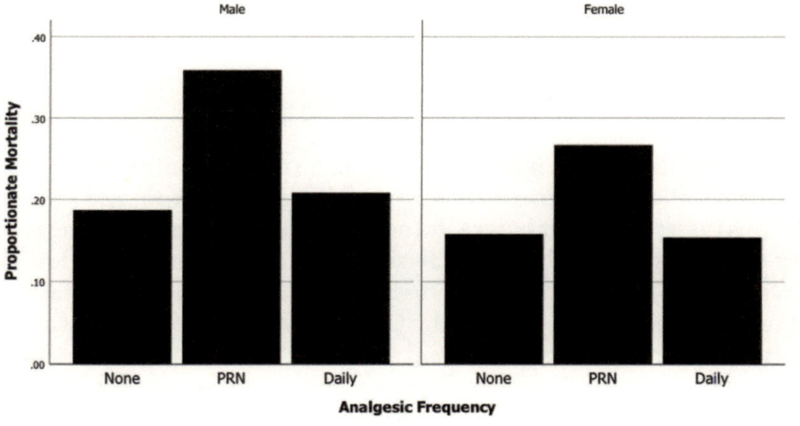

Figure 3.
Proportionate mortality by analgesic frequency in men and women.

2.2.2 Analysis of time to mortality against prescriptions for psychotropic medications

The purpose of this analysis is to advance findings from the preceding analysis in two ways. The first is to replace the binary target variable with a nominal variable

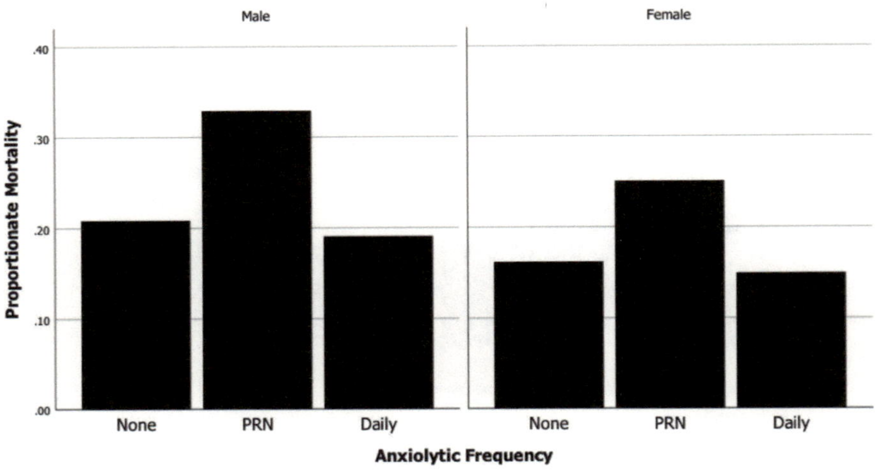

Figure 4.
Proportionate mortality by anxiolytic frequency in men and women.

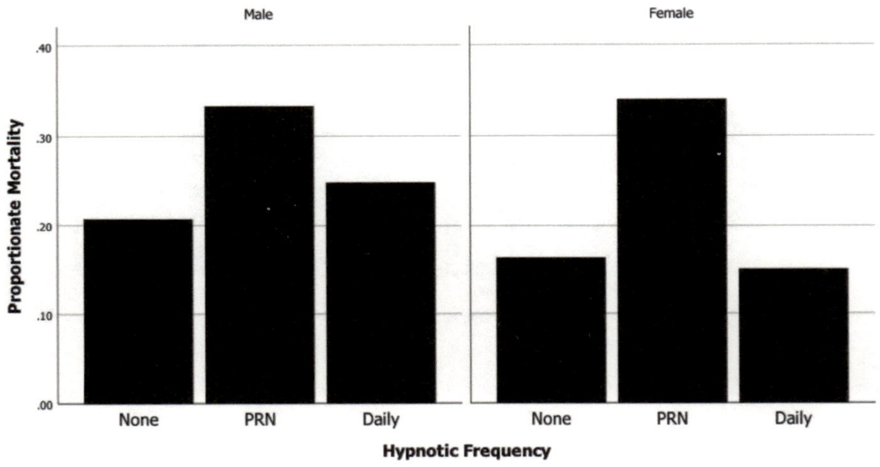

Figure 5.
Proportionate mortality by hypnotic frequency in men and women.

of time to death. The categories are 1–7, 8–30 and 31–90 days after the final assessment. We chose these categories to index mortality shortly, soon, and sometime after the final assessment but before the next scheduled assessment. The reference category is no recorded death. This analysis allows to use examine relationships between mortality and psychotropic prescriptions using a finer temporal scale.

The second advance is to include the CHESS scale [10] as a fixed effect. The CHESS is arguably the strongest current predictor of mortality for people within continuing care contexts. For example, mortality in the present database is <10% for residents with the lowest score but >85% for residents with the highest score. Consequently, inclusion of the CHESS helps us to test between two interpretations of PRN/mortality relationships. On the one hand, if nearness of death is the primary reason for PRN prescription (i.e., prescribed mainly for palliative reasons), its relationship to mortality should nullify after inclusion of the CHESS as a fixed effect. On the other hand, if PRN prescription increases the risk of subsequent mortality, a relationship should endure despite inclusion of the CHESS as a fixed effect.

The target variable in this analysis has a multinomial distribution with a generalized logit link. The random and fixed effects are the same as in the preceding analysis except for the addition of the CHESS as a fixed effect. The fixed effect coefficients for this analysis comprise a multi-layered table, wherein each layer represents a different time to death category. However, we present each layer as a separate table in order to improve ease of readability.

Tables 3–5 respectively show findings from 1–7, 8–30, 31–90 days from the final assessment. The findings for demographic, functional capability and mortality risk measures show the following trends. Over all three mortality intervals, mortality is lower in women than men, higher at older ages, higher with greater activity limitation and higher with higher CHESS scores. For the interval 8–30 days after the final assessment, mortality is lower with higher cognitive impairment (i.e., higher scores on the scale).

Compared to a reference category of no medication, the findings for prescribed medications show the following. **Table 3** shows mortality 1–7 days from the final assessment to be significantly lower with daily antidepressant prescription but significantly higher with PRN prescription of antipsychotics, analgesics and antidepressants.

Table 4 shows mortality 8–30 days after the final assessment is significantly lower with daily antidepressant and daily analgesic prescriptions but significantly higher with PRN anxiolytics and PRN analgesic prescriptions.

Model term	Odds ratio	Sig.	95% confidence interval	
			Lower	Upper
Intercept	0.017	0.000	0.013	0.022
Female	0.533	0.000	0.432	0.659
Male				
Age at assessment	1.032	0.000	1.017	1.048
Activities of daily living	1.625	0.000	1.484	1.780
Cognitive performance	0.986	1.000	0.922	1.053
CHESS scale	2.489	0.000	2.294	2.700
Daily antipsychotic	0.859	0.222	0.670	1.102
PRN antipsychotic	2.139	0.009	1.273	3.594
No antipsychotic				
Daily antidepressant	0.606	0.000	0.487	0.755
PRN antidepressant	1.696	0.045	1.010	2.847
No antidepressant				
Daily analgesic	1.104	0.395	0.863	1.410
PRN analgesic	2.655	0.000	1.916	3.681
No analgesic				
Daily anxiolytic	0.997	0.903	0.716	1.388
PRN anxiolytic	1.006	0.942	0.617	1.640
No anxiolytic				
Daily hypnotic	1.104	0.586	0.711	1.716
PRN hypnotic	1.592	0.318	0.666	3.806
No hypnotic				

Table 3.
Fixed effect odds ratios for predictors of mortality within 7 days after the final assessment.

Model term	Odds ratio	Sig.	95% confidence interval	
			Lower	Upper
Intercept	0.089	0.000	0.077	0.103
Female	0.596	0.000	0.525	0.677
Male				
Age at assessment	1.022	0.000	1.013	1.031
Activities of daily living	1.495	0.000	1.423	1.572
Cognitive performance	0.947	0.047	0.909	0.986
CHESS scale	1.883	0.000	1.786	1.986
Daily antipsychotic	0.881	0.084	0.761	1.020
PRN antipsychotic	1.335	0.280	0.872	2.043
No antipsychotic				
Daily antidepressant	0.675	0.000	0.594	0.768
PRN antidepressant	1.152	0.497	0.758	1.753
No antidepressant				
Daily analgesic	0.824	0.007	0.719	0.943
PRN analgesic	1.574	0.000	1.274	1.946
No analgesic				
Daily anxiolytic	0.885	0.337	0.721	1.086
PRN anxiolytic	1.482	0.006	1.113	1.974
No anxiolytic				
Daily hypnotic	1.049	0.624	0.802	1.370
PRN hypnotic	1.585	0.148	0.872	2.880
No hypnotic				

Table 4.
Fixed effect odds ratios for predictors of mortality 8–30 days after the final assessment.

During the 31–90 day interval after the final assessment, **Table 5** shows significantly lower mortality with daily antidepressant prescription but significantly higher mortality with PRN analgesic, anxiolytic, and hypnotic prescriptions.

At one or more levels of the target variable, the findings from this analysis replicate those from the preceding analysis that relate male gender, older age and greater activity limitation to significantly augmented mortality. They also replicate findings of augmented mortality with PRN prescription for all types of psychotropic medication. They further extend relationships of attenuated mortality to include daily prescription of both analgesic and antidepressant medication. The only failure of replication is the absence of significantly attenuated mortality with daily prescription of antipsychotic medication, despite a coefficient close to significance for the 8–30 day interval. Other significant findings include attenuated mortality with cognitive impairment (i.e., higher scores on the scale) at the 8–30 day interval and augmented mortality at each level of the target variable with higher estimates of mortality risk on the CHESS.

The inclusion of the CHESS provides a rationale to interpret findings of attenuated or augmented mortality associated with psychotropic prescriptions. The presence of such findings after control for the CHESS supports an interpretation that

Model term	Odds ratio	Sig.	95% confidence interval	
			Lower	Upper
Intercept	0.140	0.000	0.124	0.158
Female	0.650	0.000	0.587	0.720
Male				
Age at assessment	1.039	0.000	1.031	1.046
Activities of daily living	1.326	0.000	1.277	1.377
Cognitive performance	0.986	0.901	0.953	1.020
CHESS scale	1.542	0.000	1.474	1.614
Daily antipsychotic	0.989	0.807	0.882	1.108
PRN antipsychotic	1.436	0.092	1.003	2.056
No antipsychotic				
Daily antidepressant	0.822	0.000	0.743	0.909
PRN antidepressant	1.226	0.269	0.854	1.758
No antidepressant				
Daily analgesic	0.894	0.052	0.803	0.995
PRN analgesic	1.406	0.000	1.172	1.687
No analgesic				
Daily anxiolytic	1.011	0.689	0.866	1.181
PRN anxiolytic	1.392	0.006	1.093	1.772
No anxiolytic				
Daily hypnotic	1.080	0.376	0.875	1.333
PRN hypnotic	1.990	0.006	1.243	3.186
No hypnotic				

Table 5.
Fixed effect odds ratios for predictors of mortality 31–90 days after the final assessment.

the medicinal prescriptions have direct or indirect effects on mortality beyond the levels of risk measured by the tool. An alternative interpretation that the findings on mortality are secondary to altered prescribing practices with perceived closeness to death seems less plausible.

2.2.3 Analysis of predictors of multiple daily and PRN psychotropic prescriptions

LTCH residents may receive more than a single type of psychotropic medication on a daily or PRN basis. The purpose of these analyses is to identify variables that contribute to the number of such prescriptions. The target variable in both analyses is the number of psychotropic categories prescribed on a daily or PRN basis. We modeled the target as an ordinal measure with a cumulative logit link in mixed multinomial logistic regression models. Levels on the daily prescription distribution are 0, 1, 2, 3 and 4+ ordinal categories. Levels on the PRN prescription distribution are 0, 1, 2 and 3+ ordinal categories.

The fixed effects include the same demographic, functional capability and mortality risk measures as in the preceding analyses. Other measures are scales (i.e., Depression, Aggressive Behavior, and Pain scales), diagnoses (i.e., Anxiety Disorder, and Dementia) and conditions (i.e., Insomnia) that might influence psychotropic medication use. We added a final fixed effect of the total number of prescribed medications.

Model term	Odds ratio	Sig.	95% confidence interval	
			Lower	Upper
Threshold for Number = 0	0.225	0.000	0.214	0.237
Threshold for Number = 1	1.504	0.000	1.433	1.579
Threshold for Number = 2	9.442	0.000	8.970	9.939
Threshold for Number = 3	87.211	0.000	81.593	93.216
Female	1.047	0.002	1.017	1.079
Male				
Age at assessment	0.973	0.000	0.971	0.974
Activities of daily living	1.015	0.003	1.005	1.025
Cognitive performance	1.118	0.000	1.105	1.131
Depression scale	1.108	0.000	1.100	1.115
Aggressive behavior scale	1.067	0.000	1.060	1.074
Pain scale	1.208	0.000	1.187	1.230
CHESS scale	0.915	0.000	0.900	0.929
Anxiety disorder	2.744	0.000	2.599	2.896
No anxiety disorder				
Insomnia	1.124	0.000	1.079	1.170
No insomnia				
Dementia	1.393	0.000	1.348	1.439
No dementia				
Total medications	1.191	0.000	1.187	1.195

Table 6.
Fixed effect odds ratios for predictors of number of daily medications for the final assessment.

The findings in **Table 6** show fixed findings for daily prescriptions of psychotropic medication. The table includes thresholds (i.e., intercepts) at different levels of the target variable and coefficients for the predictors. Predictors of significantly higher frequencies of daily prescription include categorical measures of female gender, anxiety disorder, insomnia, and dementia; higher scores on scales measuring activity limitation, cognitive impairment, depression, aggressive behavior, and pain; and the total number of medications. These findings suggest that residents prescribed more types of psychotropic medication have higher levels on multiple conditions that might benefit from psychotropic intervention. Findings that seem at odds with the preceding are relationships of younger age and lower mortality risk (i.e., on the CHESS) to higher frequencies of daily psychotropic prescriptions.

Table 7 shows fixed effect findings for PRN prescriptions of psychotropic medication. Significant predictors of higher levels of PRN prescribing are limited to the Aggressive Behavior, Pain, and CHESS scales, and the presence of insomnia. Age and diagnosed dementia have negative relationships with the number of PRN prescriptions.

The most revealing differences in outcome of the two preceding analyses relate to diagnosed dementia and the CHESS scale. Diagnosed dementia relates positively to the number of daily but negatively to the number of PRN prescriptions. Conversely, higher mortality risk relates negatively to the number of daily but positively to the number of PRN prescriptions.

DOI: http://dx.doi.org/10.5772/intechopen.85971

Model term	Odds ratio	Sig.	95% confidence interval	
			Lower	Upper
Threshold for Number = 0	7.737	0.000	6.970	8.587
Threshold for Number = 1	73.344	0.000	63.257	85.040
Threshold for Number = 2	399.408	0.000	303.212	526.123
Female	1.033	0.496	0.941	1.134
Male				
Age at assessment	0.993	0.029	0.988	0.999
Activities of daily living	1.019	0.252	0.987	1.052
Cognitive performance	0.977	0.194	0.942	1.012
Depression scale	1.008	0.437	0.988	1.028
Aggressive behavior scale	1.085	0.000	1.064	1.107
Pain scale	1.334	0.000	1.268	1.404
CHESS scale	1.214	0.000	1.165	1.266
Anxiety disorder	0.991	0.917	0.838	1.173
No anxiety disorder				
Insomnia	1.618	0.000	1.444	1.812
No insomnia				
Dementia	0.789	0.000	0.710	0.877
No dementia				
Total medications	0.996	0.423	0.986	1.006

Table 7.
Fixed effect odds ratios for predictors of number of PRN medications for the final assessment.

2.2.4 Analysis of mortality against multiple daily and PRN psychotropic prescriptions

The purpose of this analysis is to ascertain relationships between mortality within 90 days of the final assessment and the distribution of prescriptions of psychotropic medications at the final assessment. The target variable has a binomial distribution with a complementary log-log link. The fixed effect variables include the same demographic, functional capability and mortality risk measures as previous analyses. Categories within the distribution of prescriptions include (1) no prescriptions, (2) daily and PRN prescriptions, (3) PRN prescriptions only and (4) daily prescriptions only. The reference category is zero prescriptions.

The findings in **Table 8** show significantly lower mortality for women than men and for residents with greater cognitive impairment (i.e., lower scores on the scale). Mortality is significantly higher at older ages and for residents scoring higher on activity limitation and the CHESS measure of mortality risk. Findings for the distribution of prescriptions show significantly attenuated mortality for residents with daily prescriptions only, significantly augmented mortality residents with PRN prescriptions only, and significantly augmented mortality for residents with both daily and PRN prescriptions. Paired comparisons with Bonferroni correction show significant differences across all four categories.

We interpret these findings as follows. Antipsychotics, antidepressants and analgesics have the highest rates of daily prescription. Daily prescription of each of those psychotropics attenuates mortality in at least one of our analyses. Consequently, the present association between attenuated 90-day mortality in residents with only daily prescriptions is unsurprising. For similar reasons, findings of augmented mortality in

Model term	Odds ratio	Sig.	95% confidence interval	
			Lower	Upper
Intercept	0.21	0.000	0.19	0.23
Female	0.67	0.000	0.63	0.72
Male				
Age at assessment	1.03	0.000	1.02	1.03
Activities of daily living	1.36	0.000	1.32	1.40
Cognitive performance	0.97	0.007	0.95	0.99
CHESS scale	1.58	0.000	1.53	1.62
Only daily prescriptions	0.76	0.000	0.69	0.84
Only PRN prescriptions	1.42	0.000	1.21	1.66
Daily and PRN prescriptions	1.17	0.025	1.02	1.34
No prescriptions				

Table 8.
Fixed effect odds ratios for mortality within 90 days of the final assessment.

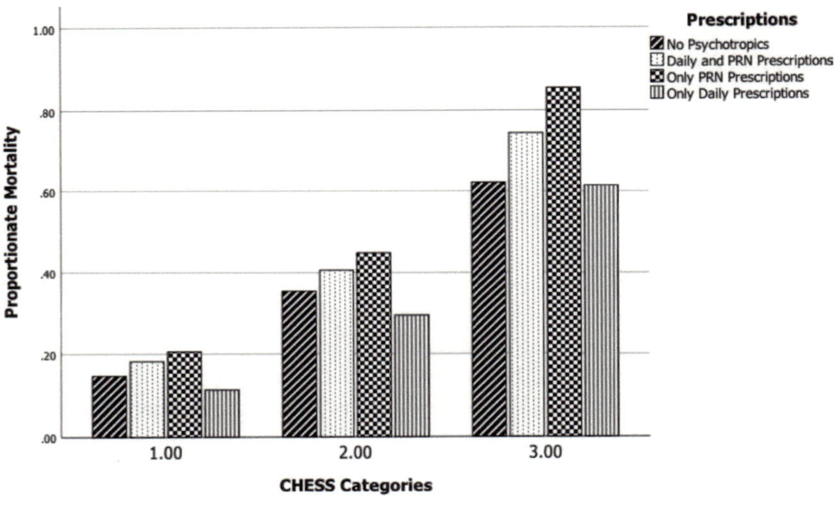

Figure 6.
Proportionate mortality by prescription regimens and CHESS categories.

residents with only PRN prescriptions are also unsurprising. The new ground breaking findings, however, concern residents with both daily *and* PRN prescriptions. This combination significantly augments mortality from levels that typify daily prescription and no prescription, but significantly attenuates levels that typify only PRN prescriptions. **Figure 6** shows that the respective mortality levels are independent of mortality risk, as estimated by categories of CHESS scores. Consequently, we infer that PRN prescription may overturn protective effects associated with daily prescriptions, whereas the latter may reduce the deleterious effects of PRN prescribing.

2.2.5 Additional analyses

Because of the statistical significance of LTCH as a random effect in all the preceding analyses, we report here on additional analyses that explore whether particular differences among LTCH contribute to fixed effect relationships to mortality. Specifically, the analyses include facility-level covariates that correspond with previously significant, resident-level fixed effect terms. These centered facility-level covariates are LTCH means for age, level of activity limitation, and mortality risk (i.e., CHESS scores); and percentages of residents by gender and psychotropic prescription on a PRN basis. The findings indicate that these facility-level variables have significant zero-order relationships with mortality in the same directions as the corresponding resident-level measure. However, all are nonsignificant when added to the list of fixed effects described in the preceding section (i.e., Section 2.2.4), with the odds-ratios and significance levels for resident-level terms showing minimal change from those in **Table 8**. We conclude, therefore, that the findings with resident-level fixed effects show negligible influence due to distributions in LTCH.

2.3 Discussion

This chapter charts a journey of scientific investigation with the following milestones:

1. Concern about excess mortality with the use of antipsychotic medication.

2. Findings that limit the preceding relationship to PRN prescription.

3. Findings that excess mortality with PRN prescription applies to all types of psychotropic medication.

4. Findings of attenuated mortality with daily prescription for some types of psychotropic medication but augmented mortality with PRN prescription for all types of medication.

5. Evidence to support the preceding after control for confounders that include a proven indicator of mortality risk.

6. Findings that diagnosed dementia relates positively to the number of daily but negatively to the number of PRN prescriptions, whereas higher mortality risk relates negatively to the number of daily but positively to the number of PRN prescriptions.

7. Evidence that a combination of daily and PRN prescriptions of psychotropics overturns protective effects against mortality associated with the former but lowers deleterious effects associated with PRN prescribing.

The journey began with concern about excess mortality with the use of antipsychotic medication but ends with evidence for minor protective effects of daily prescription. How do we account for this discrepancy? Consider, first, the quality of measurement and, second, the appropriateness of the analyses. We have confidence in the data quality and psychometric adequacy (i.e., reliability and validity) of the RAI 2.0 measures. For nearly 30 years, efforts by researchers from across the world tried to ensure good measurement quality. A recent example is research on 15 years of archived data in Canada. Those authors report favorable outcomes, concluding that the RAI 2.0 provides a "robust, high quality data source" ([28], p. 27). If our data possess unrecognized error in measurement, similar error likely contaminates findings from many hundred referred publications that rely on RAI 2.0 data.

Second, we consider the technical details of our methodology to be appropriate. The decision to restrict analysis to data from new admissions ensures that the findings from over 20,000 residents provide 1-year incidence rates (rather than prevalence rates) for mortality. Such restriction also ensures that the data from all residents begin with comparable intake assessments. Earlier in this chapter, we provide conceptual reasons to justify the appropriateness of GLMM analysis. In every analysis, findings of significant random effects for LTCH provide empirical justification. Our wish for the future is that more researchers adopt similar modeling in order to avoid incorrect assumptions that measurement error is uncorrelated in clustered data.

We also ought to mention here that the statistical models show reasonable data fit in every analysis. The final model in **Table 8** provides an example. Compared to a simplified model that includes only demographics as fixed effects, the proportions of correct classification in the full model improve by 16.3% for predicted mortality (i.e., from 50 to 66.3%) and by 2.4% for the prediction of nonmortality (i.e., from 82.1 to 84.5%). These gains are fairly impressive given the lopsided skew in mortality data.

We conclude from the preceding is that our findings are trustworthy. How then can we account for findings from other studies of excessive mortality among elderly people prescribed with antipsychotic medication? We offer three possible interpretations. The first is a failure by many or most such studies to distinguish between daily and PRN prescribing. The second is a failure to provide information about PRN prescription of other psychotropic medications. Because our findings suggest associations between augmented mortality and PRN prescription of *any* psychotropic, each of the preceding concerns anticipates overestimation of mortality.

The third interpretation relates to adherence. Anyone with work experience in LTCH settings knows that adherence to drug regimens is problematic. Residents that put pills into their mouths do not necessarily swallow them. Instead, some residents chose to hide their pills, others throw them away. Previous findings show that nonadherence to antidepressant medication has unfavorable implications for mortality [19]. Findings from psychiatry show frequent low adherence to antipsychotic medication [29]. That study reports rates of nonadherence to antipsychotic medication by psychiatric patients as high as 40% within a year. Because low adherence results in intermittent use of antipsychotic medication, the consequences in LTCH residents could well include excessive mortality.

Our scientific journey progressed past milestones that indicate consistency in findings with alternative indices of mortality and control of confounding variables. At the end of the journey, the metaphorical villain is no longer daily prescription of psychotropic medication but PRN prescription. Although relatively few residents receive PRN prescriptions on a per drug basis, one resident in eight (12.5%) has at least one such prescription. Although residents prescribed PRN medication have high mortality risk, as measured by the CHESS, **Figure 6** shows elevated mortality rates at every level of mortality risk.

Issues that arise from our research include reasons for the high mortality associated with PRN prescriptions and implications for health care practice and policy. However, before such discussion, here are some words of caution.

The present research design is correlational rather that experimental. Only the latter assigns individuals randomly to the various conditions. Despite agreement that correlational designs provide less-compelling evidence about causality than experimental designs, the majority of studies on the effects of psychotropic medications are correlational. For example, most recent studies of the effects of antipsychotic medication on mortality have correlation designs that compare residents prescribed a given drug with residents without prescription or prescribed an alternative medication [6]. However, even in experimental studies, which are invariably of short duration [5], any differences in mortality between conditions do not necessarily imply direct causation. Such effects may arise because of changes to health stability or changes in likelihood of adverse health events. We make these points to warn against simplistic interpretation of complex phenomena.

Efforts within epidemiology to facilitate appropriate causal inferences include procedures to identify *confounding by indication*, which "refers to a situation where patient characteristics, rather than the intervention, are independent predictors of outcome" ([30], p. 841). The present analyses encompass procedures used to evaluate confounding by indication, which include, but are not limited to, the following. Covariate adjustment in which potential confounders are added to the list of fixed effects in modeling (e.g., age, gender, and functional measures). "Propensity" scores were used, which relate to the likelihood of intervention (e.g., high CHESS scores increase the likelihood of PRN prescription). The use of instrumental variables presumed to substitute for unmeasured confounders (e.g., LTCH and properties of LTCH as random and fixed effect variables, respectively). Although other research reports differences in causal direction depending on the kind of analysis [30], the present findings indicate directional consistency in relationships between mortality and PRN prescribing before (**Figures 1–5**) and after adjustment of variables from all three classes of confound.

However, our analyses clearly omit control for unmeasured sources of confounding. Also, our research lacks the rigor that random assignment brings to experimental forms of design. Consequently, we strongly recommend that future randomized controlled trials evaluate the relationships we report here.

Our cautious interpretation of effects on mortality associated with daily and PRN prescription of psychotropic medication rests heavily on findings from the final analysis. These findings indicate that PRN prescription overturns any protective effects associated with daily prescription, whereas daily prescription lowers adverse consequences of PRN prescribing. These findings suggest a mutually antagonistic relationship between daily and PRN prescriptions with respect to their effects on mortality. The clinical rationale for daily prescription is the alleviation of disturbance to a resident's equilibrium, as exemplified by aggression, depression, pain, anxiety and insomnia. Successful treatment results in regained equilibrium after the resident adapts to regular ingestion of the drug. Conversely, the reasoning behind PRN prescription is to alleviate disturbances to equilibrium thought to be temporary. However, irregularity of ingestion is antagonistic to adaptation and may exacerbate instability through toxicity rather than help restore equilibrium. This interpretation appears to be consistent with our findings. Practical implications for health care practices and policy are consistent with current trends. Such trends include the de-prescribing of psychotropic medications for older people, with such medications to be used only for short durations and as a last resort [5, 6]. Similarly, a recent rewrite of Medicaid Long Term Care rules in the USA states that LTCH residents must not be prescribed antipsychotic, antidepressant, anxiolytic or hypnotic medications unless necessary to treat a specific condition diagnosed and documented in the clinical record. The new rules also normally limit PRN prescriptions to 14 days. Any extension requires an attending physician or prescribing practitioner to document the rationale in the resident's medical record [31]. We especially applaud these latter rules, which tackle an issue hitherto neglected in health care practice and policy.

3. Conclusion

PRN prescriptions of psychotropic medications have more aversive effects on the mortality of LTCH residents than daily prescriptions. These findings apply to all types of psychotropic medication after control for major confounding variables. The findings also suggest that daily prescription may ameliorate effects on mortality associated the PRN prescriptions, whereas PRN prescriptions exacerbate effects on mortality associated with daily prescription.

Acknowledgements

All the authors contributed to the research and manuscript preparation, and verified their authorship of this chapter. The authors wish to thank the editors, Robert Reynolds and Steven Day, for very helpful comments on an earlier draft.

Conflict of interest

No author has any conflict of interest.

Author details

Michael J. Stones*, Sarah Worobetz, Jason Randle, Carlina Marchese, Shauna Fossum, Dane Ostrom and Peter Brink
Lakehead University, Thunder Bay, Ontario, Canada

*Address all correspondence to: mstones@lakeheadu.ca

References

[1] Kozman MN, Wattis J, Curran S. Pharmacological management of behavioural and psychological disturbance in dementia. Human Psychopharmacology: Clinical and Experimental. 2006;**21**(1):1-2. DOI: 10.1002/hup.745

[2] Cerejeira J, Lagarto L, Mukaetova-Ladinska E. Behavioral and psychological symptoms of dementia. Frontiers in Neurology. 2012;**3**:73. DOI: 10.3389/fneur.2012.00073

[3] Kapur S, Mamo D. Half a century of antipsychotics and still a central role for dopamine D2 receptors. Progress in Neuro-Psychopharmacology and Biological Psychiatry. 2003;**27**(7):1081-1090. DOI: 10.1016/j.pnpbp.2003.09.004

[4] Prakash S, Masand MD. Side effects of antipsychotics in the elderly. The Journal of Clinical Psychiatry. 2000;**61**:43-49

[5] Banerjee S. The Use of Antipsychotic Medication for People with Dementia: Time for Action. London: Department of Health, UK Government; 2009. Available from:http://psychrights.org/research/digest/nlps/banerjeereportongeriatricRalph SJ, Espinet AJ. Increased all-cause mortality by antipsychotic drugs: Updated review and meta-analysis in dementia and general mental health care. Journal of Alzheimer's Disease Reports. 2018;**2**(1):287-312. DOI: 10.3233/ADR-170042

[6] Brink P, Stewart S, Stones MJ. Institutional reactions to verbally abusive behaviour by residents of long-term care facilities. In: Proceedings of the Annual Conference of the Ontario Network for the Prevention of Elder Abuse. Toronto, ON; 2005

[7] Jones RN, Marcantonio ER, Rabinowitz T. Prevalence and correlates of recognized depression in US nursing homes. Journal of the American Geriatrics Society. 2003;**51**(10):1404-1409. DOI: 10.1046/j.1532-5415.2003.51458.x

[8] Neufeld E, Freeman S, Joling K, Hirdes JP. "When the golden years are blue" Changes in depressive symptoms over time among older adults newly admitted to long-term care facilities. Clinical Gerontologist. 2014;**37**(3):298-315. DOI: 10.1080/07317115.2014.885919

[9] Hirdes JP, Poss JW, Mitchell L, Korngut L, Heckman G. Use of the interRAI CHESS scale to predict mortality among persons with neurological conditions in three care settings. PLoS One. 2014;**9**(6):e99066. DOI: 10.1371/journal.pone.0099066

[10] Worobetz S. Effects of antipsychotic medications on older adults with dementia in Canadian complex and long-term care facilities [thesis]. Thunder Bay: Lakehead University; 2014

[11] Stasinopoulos J, Bell JS, Ryan-Atwood TE, Tan EC, Ilomäki J, Cooper T, et al. Frequency of and factors related to pro re nata (PRN) medication use in aged care services. Research in Social and Administrative Pharmacy. 2018;**14**(10):964-967. DOI: 10.1016/j.sapharm.2017.11.004

[12] La Frenais FL, Bedder R, Vickerstaff V, Stone P, Sampson EL. Temporal trends in analgesic use in long-term care facilities: A systematic review of international prescribing. Journal of the American Geriatrics Society. 2018;**66**(2):376-382. DOI: 10.1111/jgs.15238

[13] Holmerová I, Auer SR, Beránková A, Höfler M, Ratajczak P, Šteffl M. Cognitive status and use of analgesics and anxiolytics in residents of nursing homes in the Czech Republic. Clinical Interventions in Aging. 2018;**13**:2511. DOI: 10.2147/CIA.S18860

[14] Bhattacharjee S, Oh YM, Reiman EM, Burke WJ. Prevalence, patterns, and predictors of depression treatment of community-dwelling elderly individuals with dementia in the United States. The American Journal of Geriatric Psychiatry. 2017;**25**(7):803-813. DOI: 10.1016/j.jagp.2017.03.003

[15] Enache D, Fereshtehnejad SM, Kareholt I, Cermakova C, et al. Antidepressants and mortality risk in a dementia cohort: Data from SveDem, the Swedish Dementia Registry. Acta Psychiatrica Scandinavica. 2016;**134**: 430-440. DOI: 10.1111/acps.12630

[16] Zivin K, Kim HM, Yosef M, Maust DT, et al. Antidepressant medication treatment and risk of death. Journal of Clinical Psychopharmacology. 2016;**36**(5):445-452. DOI: 10.1097/ JCP.0000000000000545

[17] Bali V, Chatterjee S, Johnson ML, Chen H, et al. Risk of mortality in elderly nursing home patients with depression using paroxetine. Pharamcotherapy. 2017;**37**(3):287-296. DOI: 10.1007/s11920-018-

[18] Krivoy A, Stubbs B, Balicer RD, Weizman S, et al. Low adherence to antidepressants is associated with increased mortality risk: A large nationally representative cohort study. European Neuropsychopharmacology. 2017;**27**:970-976

[19] Biffi A, Scotti L, Rea F, Lucenteforte E, et al. Adherence to antidepressants and mortality in elderly patients with cardiovascular disease. Clinical Drug Investigation. 2018;**38**:593-602. DOI: 10.1080/17425255.2019.1561860

[20] Andreescu C, Varon D. New research on anxiety disorders in the elderly and an update on evidence-based treatments. Current Psychiatry Reports. 2015;**17**:53. DOI: 10.1007/ s11920-015-0595-8

[21] Hwang Y, Chen T, Wen Y, Liou H. Does the prescription of anxiolytic and hypnotic drugs increase mortality in older adults? Journal of the American Geriatrics Society. 2015;**63**(6):1263-1265. DOI: 10.1111/jgs.13466

[22] Weich S, Pearce HL, Croft P, Singh S, Crome I, Bashford J, et al. Effect of anxiolytic and hypnotic drug prescriptions on mortality hazards: Retrospective cohort study. BMJ. 2014;**2014**:348. DOI: 10.1136/bmj.g1996

[23] Usher K, Holmes C, Lindsay D, Luck L. PRN psychotropic medications: The need for nursing research. Contemporary Nurse. 2003;**14**(3):248-257. DOI: 10.5172/conu.14.3.248

[24] Lindsey PL. Psychotropic medication use among older adults: What all nurses need to know. Journal of Gerontological Nursing. 2009;**35**(9):28-38. DOI: 10.3928/00989134-20090731-01

[25] Stokes JA, Purdie DM, Roberts MS. Factors influencing PRN medication use in nursing homes. Pharmacy World & Science. 2004;**26**(3):148-154. DOI: 10.1023/B:PHAR.0000026803.89436.a8

[26] Wright S, Stewart D, Bowers L. Psychotropic PRN Medication in Inpatient Psychiatric Care: A Literature Review. London: Section of Mental Health Nursing, Institute of Psychiatry; 2012. Available from: https://www.kcl.ac.uk/ioppn/depts/hspr/archive/mhn/projects/PRN-Medication

[27] Hirdes J, Poss JW, Caldarelli H, Fries BE, Morris JN, Teare GF, et al. An evaluation of data quality in Canada's continuing care reporting system (CCRS): Secondary analyses of Ontario data submitted between 1996 and 2011. BMC Medical Informatics and Decision Making. 2013;**13**:27. Available from: http://www.biomedcentral.com/1472-6947/13/27

[28] Perkins DO. Adherence to antipsychotic medications. The Journal of Clinical Psychiatry. 1999;**60**(Suppl 21): 25-30. Available from: https://europepmc.org/abstract/med/10548139s

[29] Cnossen MC, van Essen TA, Ceyisakar IE, Polinder S, Andriessen TM, van der Naalt J, et al. Adjusting for confounding by indication in observational studies: A case study in traumatic brain injury. Clinical Epidemiology. 2018;**10**:841-852. DOI: 10.2147/CLEP.S154500

[30] Barlas S. Medicare adds new long-term-care pharmacy rules: Agency passes again on pharmacist Independence requirements. Pharmacy and Therapeutics. 2016;**41**(12):762-764

Chapter 2

Evidence for the Effectiveness of Soy in Aging and Improving Quality of Life

Bahram Herman Arjmandi and Elizabeth Marie Foley

Abstract

Soy is a highly nutritious yet widely underutilized food. Because of the controversy surrounding soy, individuals with chronic disease states that may benefit from soy or soy isoflavone consumption may avoid this food. The relationship of soy to estrogen, breast cancer, osteoarthritis, and other chronic disease states is discussed. Osteoarthritis is a specific focus, as the immobility brought about by this disease state may lead to other chronic diseases that are also positively affected by soy consumption, and because there is no clear etiology or cure for this debilitating disease. Conclusions and future directions for soy research as it relates to healthy aging are also discussed.

Keywords: soy, osteoarthritis, aging, breast cancer, longevity

1. Introduction

Globally, life expectancy has increased by nearly 20 years in both sexes since the 1950s [1]. In the United States (US) in 2015, life expectancy at birth was calculated to be almost 79 years old for both males and females [2]. While these numbers are encouraging, the quality of life of these individuals has not increased along with this increased life expectancy [3]. There are many factors that can influence quality of life but chronic diseases such as osteoporosis, osteoarthritis, heart disease, sarcopenia, type 2 diabetes (T2D), and dementia all play a role in the quality of life (QOL) of aging individuals.

Many chronic diseases are highly preventable and are generally treatable through diet and exercise. Indeed, poor diet and inadequate physical activity are two of the three most common risk factors for several chronic diseases, and addressing these factors in addition to the third risk factor, smoking, reduces the risk of cardiovascular disease (CVD), stroke, and T2D by 80% [4]. A 2013 study which analyzed the effect of physical inactivity on chronic disease estimated that, worldwide, physical inactivity is linked to 6–10% of chronic diseases that included CVD, T2D, breast cancer, and colon cancer, and that inactivity is associated 9% of premature deaths [5].

Knee osteoarthritis (OA) has been ranked as one of the top contributors to global disabilities in the world [6]. Osteoarthritis is a degenerative disorder of synovial joints characterized by focal loss of articular cartilage with reactive changes in subchondral and marginal bone, synovium and para-articular structures [7]. These degenerative changes lead to the primary complaints of pain with movement, stiffness, instability, and loss of function, particularly in those with knee OA [8].

The World Health Organization (WHO) estimates that about 10% of individuals 60 years or older have OA, an estimate that will only increase as the world's population continues to age due to longer life expectancies [9]. The conclusive etiology of this disease is unknown, but injury to the joint, age, gender, and obesity are all known factors to contribute to the development of OA [10]. There is also mounting evidence that leptin may play a key role in the pathophysiology of OA. Leptin concentration in the serum is positively correlated with Body Mass Index (BMI) [11, 12]. This finding is significant as it helps to explain why obesity is a risk factor for OA, even in non-weight bearing joints such as hands.

Because individuals with OA are in constant pain, they are likely to stop exercising or to engage in any physical activity, thus increasing their risk of morbidity. It may also lead to other chronic diseases, both as a result of the lack of exercise, and the possibility of weight gain and the risks associated with excess weight. In fact, T2D has been shown to be a risk factor for knee OA progression [13], indicating that these disease states feed off of each other. While exercise is incredibly important for health, nutrition may be a much more helpful and significant treatment for individuals with chronic diseases, specifically OA, because the source of many of these diseases is underlying inflammation [14] and treating the inflammation through dietary change may result in the treatment of multiple disease states.

Although OA affects a large number of Americans, there are no proven therapies for preventing or halting its progression. In the normal joint, there is a balance between synthesis and degradation of cartilage. In inflammatory conditions such as OA, and other chronic diseases, catabolic molecules are upregulated, thereby interrupting the function of anabolic molecules [15]. Catabolic cytokines also induce the production of specific matrix degrading metalloproteases, causing cartilage degradation [16]. This finding has been confirmed by the increased level of these cytokines in people with OA [17]. Unregulated or excess production of these molecules may play a detrimental role in the pathophysiology of OA [16, 18].

The development of OA is also accompanied by increased production of prostaglandins (PGs), molecules that may contribute to joint damage, pain, and inflammation [19]. Cyclooxygenase (COX) is responsible for the production of PGs and exists as two distinct isoforms, COX-1, and COX-2. Increased expression of COX-2 has been demonstrated in synovial tissues suggesting that COX-2 expression mediates the inflammatory response in OA [20]. COX-2 is undetectable in most tissues, but is increased in inflammation leading to overproduction of PGE2 [21, 22]. Inhibition of these enzymes by nonsteroidal anti-inflammatory drugs (NSAID) and selective COX-2 inhibitors reduces the levels of PGs, resulting in a reduction in pain and inflammation.

Finding nutritional interventions to target the COX-2 pathway while allowing other necessary inflammatory pathways to function would significantly increase quality of life as well as functionality of individuals with OA. It may also inadvertently target unregulated inflammation that has been associated with other chronic disease states, and allow for affected individuals to exercise thus further decreasing their risk for the aforementioned chronic diseases.

Soy appears to be a promising treatment for those with OA, and has many other health benefits. Soy protein is low in saturated fat, contains all of the essential amino acids, and is also a good source of fiber, iron, calcium, zinc, and B vitamins [23]. This book chapter will focus on soy and its relationship to OA and other chronic diseases.

2. Nutrition profile of soy

Soy is a very nutritious plant, and the only complete plant protein. Protein in soy is not only high, but comparable in quality to animal protein regarding amino acid content and digestibility [24]. The carbohydrate content of soybeans is not only low, but poorly digested by intestinal enzymes, and thus behaves as a prebiotic for beneficial bacteria [25]. The fat content is highly variable among different soybean varieties, but includes 10–15% saturated fat, 19–41% monounsaturated fat, and 46–62% polyunsaturated fat [26].

Most notably, soybeans contain isoflavones. The three main isoflavones present in soybeans include genistein (50% of isoflavones), daidzein (40% isoflavones) and glycitein (10% of isoflavones) [27]. Isoflavones are also classified as phytoestrogens because of their similar structure to estrogen (**Figures 1** and **2**). Isoflavones are more bioactive in their unconjugated (aglycon) form than their conjugated form, which must be hydrolyzed in the intestine to release the aglycons [28]. Additionally, fermented soy has more unconjugated isoflavones, thus making fermented soy foods more pharmacokinetically beneficial [29]. Soy isoflavones are also metabolized by gut bacteria, which leads to many different metabolites, the most biologically active being euqol [30]. Equol is structurally similar to estrogen, but inhibits growth of mammary tumors and may act as a selective estrogen receptor modulator (SERM) [31]. Isoflavones have anti-oxidative and anti-inflammatory properties, as well as the ability to alter gene expression, specifically in estrogen-responsive genes [32]. It is this ability that often leads health practitioners to believe that soy may be dangerous for certain populations, specifically breast cancer, which will be later discussed in this chapter. However, these SERM like capabilities are responsible for many of soy's positive effects on health.

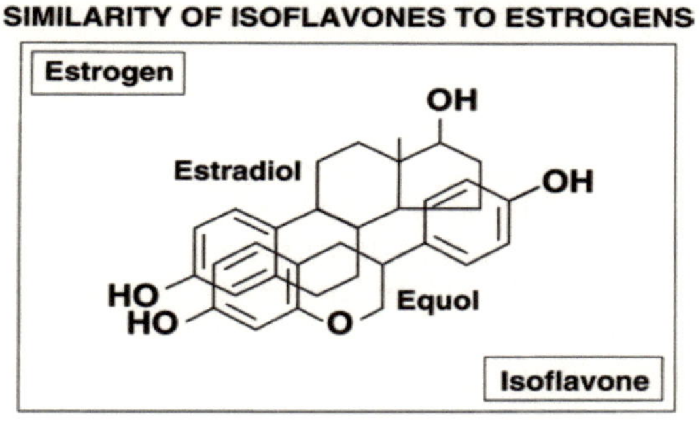

Figure 1.
Similarity of isoflavones to estrogens.

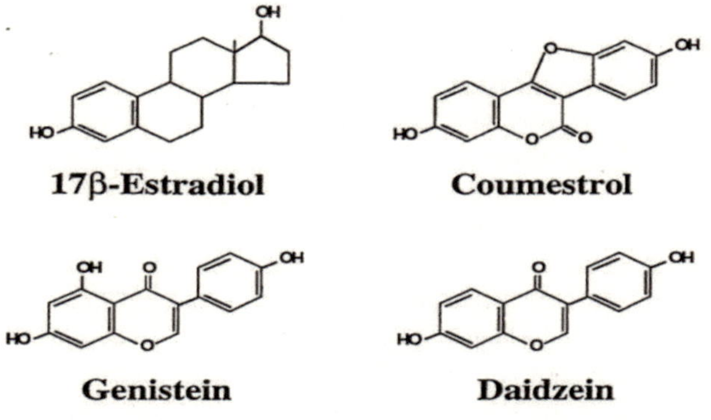

Figure 2.
Structure of estrogen and isoflavones.

3. Soy, estrogen, and breast cancer

Breast cancer is one of the most common cancers diagnosed in women in the United States, and is the second leading cause of death after lung cancer in women [33]. Breast cancer is strongly linked to ovarian hormones and estrogen levels [34]. Factors like high endogenous estrogen levels and hormone therapy have been implicated in increasing breast cancer risk [35]. Indeed, 2/3 of breast cancer cases are estrogen receptor (ER) positive [36].

Because soy isoflavones closely resemble estrogen, many health practitioners warn against soy consumption in women, women with breast cancer, and post-menopausal women for fear that soy will behave like an estrogen molecule. In our opinion, this idea is misconceived, as soy isoflavones would likely compete with endogenous estrogen for binding receptor sites in the breast, thereby reducing estrogen-stimulated growth and proliferation in the breast tissue, and may reduce endogenous estrogen concentrations [37]. Indeed, it has been shown that soy isoflavones may act as an estrogen antagonist in estrogen rich environments, and an estrogen agonist in low estrogen environments [38]; there is also evidence that the bioavailability of soy isoflavones may be inversely related to estrogen levels [39].

Epidemiological studies have shown that soy isoflavones do exert a protective effect on breast cancer risk, indicating a 16% risk reduction per 10 mg of daily isoflavone consumed [40]. A Dutch study [41] found that high levels of plasma genistein were associated with up to a 32% decreased risk of breast cancer. A 2009 study [42] that investigated soy food intake and breast cancer survival found that soy food consumption was associated with a marked decreased risk of both mortality and recurrence of breast cancer.

A 1997 study [43] found that genistein is a potent estrogen agonist and exhibited cell growth-inhibitory actions in breast cancer cells *in vitro*. A more recent study [44] also found that genistein works to inhibit topoisomerase II activity, thus resulting in the inhibition of breast cancer growth. Davis et al. [45] investigated the radioprotective effects of genistein by injecting female mice with the isoflavone 24 h prior to receiving a toxic dose of radiation, and found that genistein treated mice expressed fewer DNA damage responsive and cell cycle checkpoint genes than untreated mice. Magee et al. [46] investigated the effect of coumestrol, glycitein, daidzein, and the metabolites equol and O-desmethylangolensin on MDA-MB-231 cells, finding that each inhibited invasion by approximately 30% at the lowest dose, while genistein and coumestrol exerted the most potent inhibitory effects on invasion at the highest dose.

A clinical trial by Shike et al. [47] supplementing soy isoflavones in women with breast cancer found that soy consumption did alter gene expression in breast cancer tumors, specifically in FANCC and UGT2A1 which have both been implicated in the development of breast cancer tumors. There was a subset of tumors with upregulated FGFR2 expression, which is a marker of poor prognosis in breast cancer patients, and overall soy intake did not significantly change cell proliferation and apoptosis indices compared with the placebo group. While this initially sounds discouraging, the article points out that the clinical ramifications of this minor upregulation have yet to be determined.

Another common concern about soy supplementation in post-menopausal women, specifically, is that it causes lymphocytopenia, which is the condition of having low levels of lymphocytes in the blood. Some of these concerns stem from a multicenter study [48] published in 2001 where postmenopausal women supplemented 600 mg of ipriflavone, a synthetic isoflavone, for 3 years. Out of 234 women, 13.2% developed subclinical lymphocytopenia ($<0.5 \times 10^3/mm^3$). Another 2 year study [49] found that 3% of their participants also developed abnormal lymphocyte numbers. Another study by Ben-Hurt et al. [50] found that postmenopausal women also had higher monocyte levels, indicating that menopause definitively alters hematological parameters.

A rat study [51] by our lab refutes these results. Our study not only found that ovariectomy increased lymphocyte, monocyte, eosinophil, and basophil differential counts, but that soy isoflavones retuned leukocyte counts to pre-surgery levels. To test the truth of this in human populations, our lab also investigated the extent to which 1 year of 25 g soy protein containing 60 mg isoflavones supplementation alters lymphocyte counts in postmenopausal women [52]. This study indicated no effect on total and differential white blood cell counts in postmenopausal women, which may be due to the fact that the estimated isoflavone content of the soy protein was lower than the pharmacological dose at 60 mg.

Because leukocyte counts tend to go up with menopause, it is not necessarily a bad side effect for pharmaceutical doses of soy to bring down white blood cell counts. Additionally, the supplementation of soy protein did not have a significant impact on leukocyte levels, indicating that soy supplementation is generally safe for healthy postmenopausal populations.

4. Soy, estrogen, and OA

Interestingly, OA is often seen in postmenopausal women, and is three times more likely to be a problem for postmenopausal women rather than men [53]. Cartilage is an estrogen sensitive tissue, which may, in part, explain the gender disparity. Because postmenopausal women experience a severe drop in the production of estrogen, it stands to reason that estrogen may be protective against the development of OA. Some studies [54–56] have found an association between hysterectomy and OA, while others [57, 58] have found no association. A study by Gao et al. [59] found that estradiol (E_2) deficiency as well as changes in estrogen metabolites are involved in the pathogenesis of OA. Increased cartilage and bone turnover has been found in multiple animal models of menopause [60], but contrary to a general belief that lack of estrogen in women is the cause of OA, Tsai and colleagues [61] have suggested that excessive level of synovial fluid estrogen is responsible for the development of OA in both men and women. Indeed, some studies have found that the direct administration of estrogen to the knee joint has increased OA instance and progression [62, 63]. Intraarticular estradiol injection to ovariectomized rabbits both upregulated ER and ultimately caused further cartilage degeneration [64].

Soy isoflavones are often referred to as phytoestrogens, and may be helpful in relieving some symptoms of OA, and possibly prevent its progression. The conformational binding of soy isoflavones is similar to that of a SERM, which have been shown to be effective estrogen agonists or antagonists [65]. Genistein is the most potent of the isoflavones, and can therefore hypothetically produce positive effects on cartilage by blocking the action of estrogen. In addition to the possibility of modulating ERs, soy isoflavones may be able to increase IGF-1 production and decrease inflammation while also acting as an antioxidant. IGF-1 is thought to slow cartilage degradation [66]. Because soy isoflavones may serve as a natural modulator of IGF-1 production, it is probable that consumption of soy would benefit people suffering from OA.

5. Soy, leptin, and OA

Leptin is of particular interest in the pathology of OA, as the severity of OA is associated with both weight and BMI [67, 68], and leptin is generally elevated in obese individuals [69]. Leptin is a hormone secreted by adipocytes and is involved with energy homeostasis, namely through its ability to cross the blood brain barrier to decrease orexigenic neuropeptides and increase anorexigenic neuropeptides [70]. In healthy individuals, leptin is secreted in proportion adipose tissue and energy intake [71]. Leptin is generally thought of as a satiety hormone, although many obese individuals have "leptin resistance" [72] where the secretion of leptin in these individuals does not suppress appetite or lead to reduced energy intake.

The role of leptin may extend beyond energy homeostasis. BMI and plasma leptin levels in OA patients correlate positively [70]. Plasma leptin concentrations have also been found to be 3 times higher in premenopausal women than men [73]. Bao et al. [74] found that injecting the knee with leptin caused significant degradation of the cartilage. Additionally, leptin taken from the synovial joint has been found to be higher than plasma leptin concentrations [75].

Results from our research group, corroborates previous findings [76]. In this study, we examined the relationship between serum and synovial fluid concentrations of leptin in both males and females with varying degrees of OA. Serum and synovial fluid samples were obtained from 20 men (mean age = 68.4 ± 10.8 years) and 20 women (mean age = 61.6 ± 12.4 years) with varying degrees of OA who underwent arthroscopic or total knee replacement surgery. We found that leptin concentrations in both the serum and synovial fluid of patients with knee OA increased according to disease severity. That is, as the level of OA became more severe, the leptin concentration also increased, in both men (**Figure 3A**) and women (**Figure 3B**). We also found a significant correlation between serum and synovial fluid leptin concentration and BMI in both men (**Figure 4A** and **B**) and women (**Figure 5A** and **B**) with OA. These findings indicate that leptin may in part play a role in the increased risk of OA related to obesity.

The mechanism by which leptin may contribute to the pathophysiology of OA is likely due to its place in the cytokine family [72]. Leptin may trigger immune responses by increasing the expression of adhesion molecules, likely through a pro-inflammatory cytokine pathway [77]. Additionally, mice without a working leptin gene (ob/ob) demonstrated decreased secretion of inflammatory cytokines, while the administration of leptin to these mice restored inflammatory secretions [78]. Additionally, leptin receptors are present in the cartilage suggesting a direct action on this tissue. There is evidence [79] that leptin stimulates inflammatory markers Interlukin-6 (IL-6), Interlukin-8 (IL-8), nitric oxide, Interlukin-1 β (IL-1β), Tumor Necrosis Factor-alpha (TNFα), COX2, and PGE 2 in the joint thereby contributing to cartilage matrix breakdown.

Because of isoflavones' role in inflammation, the negative action of excess leptin levels on cartilage may be suppressed by isoflavones. For example, rats fed a high fat

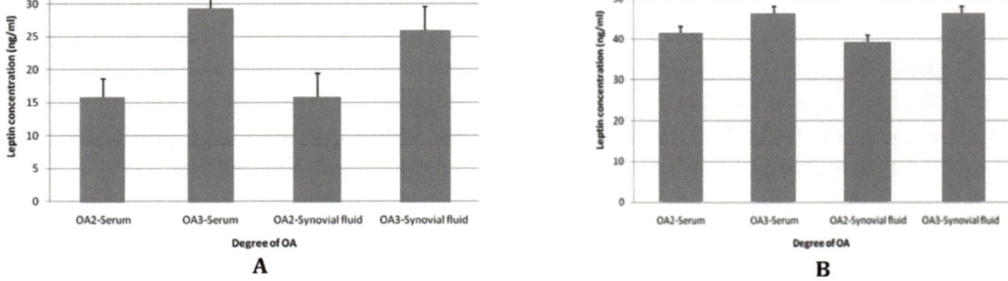

Figure 3.
The relationship between serum and synovial fluid concentrations of leptin and severity of OA in both men (A) and women (B).

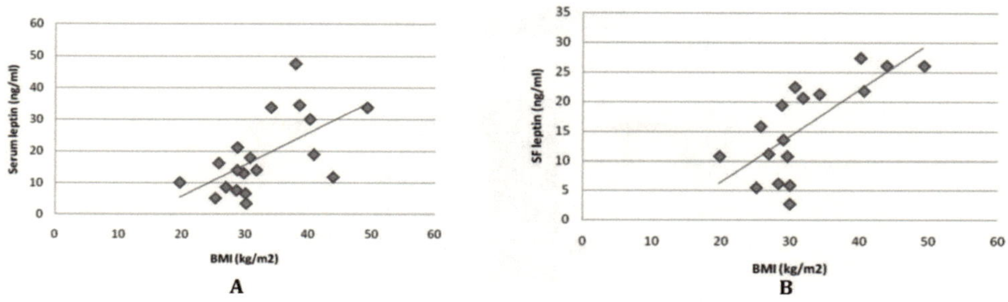

Figure 4.
The correlation between serum (A) and synovial fluid (B) leptin concentration and BMI in men with OA.

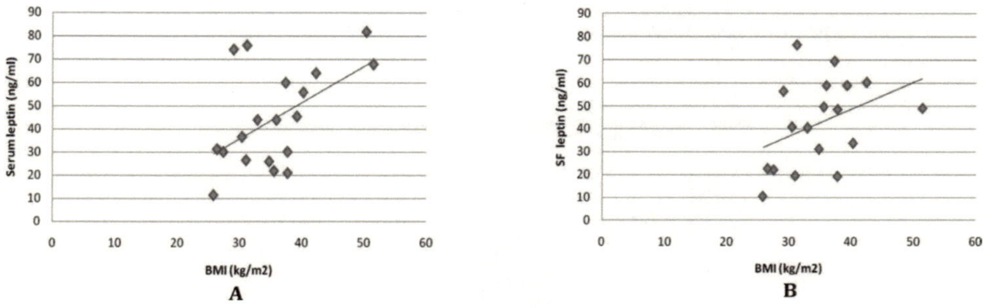

Figure 5.
The correlation between serum (A) and synovial fluid (B) leptin concentration and BMI in women with OA.

soy diet, or regular soy diet, were found to have lower serum leptin concentrations than those fed a high fat casein, or regular fat casein diet [80]. Their study [80] also found that the expression of inflammatory genes decreased along with the expression of leptin. Niwa et al. [81] also found that soy isoflavones decreased leptin secretion in the adipocytes of mice, and a study by Llaneza et al. [82] found that the consumption of 200 mg of soy isoflavone extract in postmenopausal women resulted in decreased leptin levels, as well as TNFα. Another study in overweight and obese subjects found that after 12 weeks of black soy peptide supplementation, serum leptin concentrations were significantly reduced from baseline [83].

These studies and our observations so far suggest that soy and its isoflavones are likely very efficacious in alleviating pain associated with OA and its symptoms, in part due to its ability to decrease inflammatory responses. Soy's ability to mediate leptin and inflammatory immune responses may also be integral in both preventing OA, halting its progression, and improving the QOL of individuals affected.

6. Soy and OA

The main soy isoflavones include genistein, daidzein, and glycetine [84]. Genistein is structurally similar to ipriflavone [84], a synthetic isoflavone. SERMs such as tamoxifen [85] and ipriflavone [86] have both been shown to influence cartilage metabolism and reduce or alleviate the symptoms associated with OA. Therefore, it is conceivable to also expect that genistein similarly influences cartilage metabolism.

Our *in vitro* study [87] found that genistein had the capacity to reduce inflammation in human chondrocytes. Indeed, in chondrocytes treated with LPS to induce inflammation, genistein significantly decreased COX-2 production (**Figure 6**), but

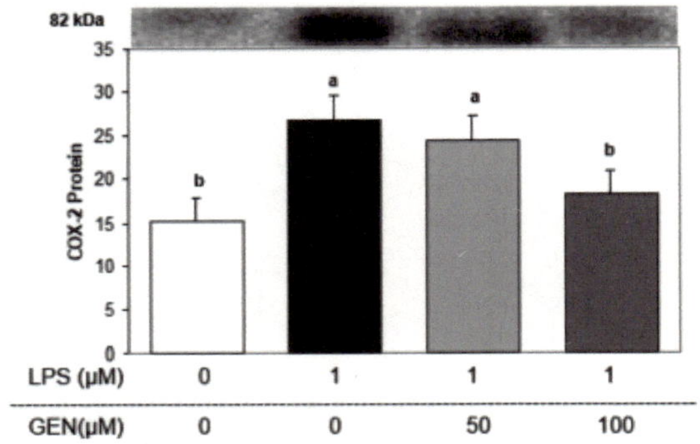

Figure 6.
COX-2 levels in cytosolic fraction of chondrocytes. LPS, lipopolysaccharides; and GEN, genistein. Bars represent mean ± SE, n = 3 per treatment group. Bars with different letters are significantly different from each other (P < 0.05).

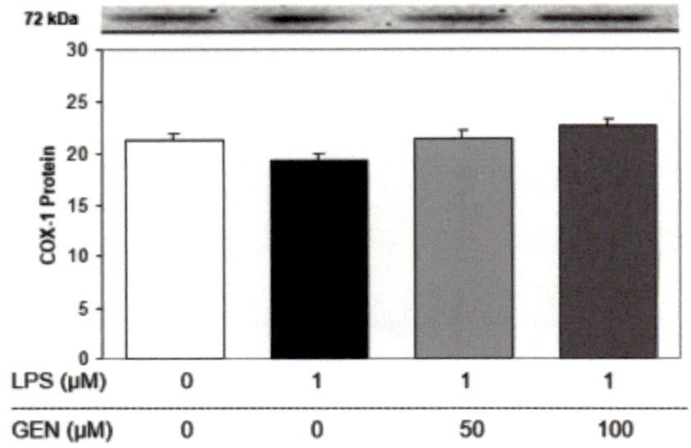

Figure 7.
COX-1 levels in cytosolic fraction of chondrocytes. LPS, lipopolysaccharides; and GEN, genistein. Bars represent mean ± SE, n = 3 per treatment group.

did not have an effect on COX-1 production (**Figure 7**) [87]. This is of particular interest, as NSAIDs are thought to inhibit inflammation via COX-1 and COX-2 dependent pathways, but are thought to cause damage because of the inhibition of COX-1, an important enzyme that regulates normal cellular processes and is expressed in most tissues [88]. This inhibited synthesis caused by most NSAIDS can negatively affect the maintenance and integrity of the gastric and duodenal mucosa, as well as lead to kidney issues [89, 90]. COX-2, however, is generally unexpressed by most tissues and is upregulated specifically by inflammation [91]. The seemingly selective inhibition of COX-2 by genistein provides a promising alternative to those who experience gastric distress due to the use of NSAIDs.

IL-1β, an inflammatory cytokine, was also measured in this study and was found to be lower in both the high and low doses of genistein (**Figure 8**) [87]. More importantly, YKL-40, a marker of human cartilage glycoprotein degradation [92], was found to be suppressed in genistein treated groups (**Figure 9**); however, the difference between the LPS and genistein groups did not reach statistical significance [87].

An animal study by Borzan et al. [93] also supports our clinical findings on soy. The aim of the aforementioned study was to determine if a soy diet could reduce the pain behaviors and inflammation induced by the intraplantar administration of complete Freund's adjuvant. They reported that neuropathic pain following partial sciatic nerve injury was attenuated in rats fed a soy protein diet [93], indicating that soy may be effective in attenuating pain symptoms, including those of OA.

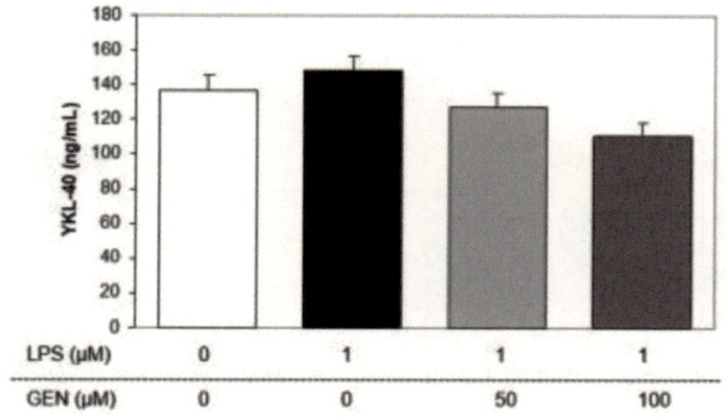

Figure 8.
IL-1β level in culture supernatant measured via ELISA kit. LPS, lipopolysaccharides; and GEN, genistein. Bars represent mean ± SE, n = 4 per treatment group.

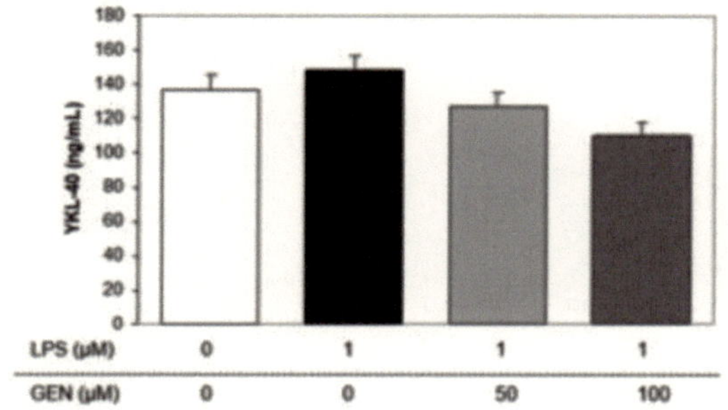

Figure 9.
YKL-40 level in culture supernatant which was measured via ELISA kit. LPS, lipopolysaccharides; and GEN, genistein. Bars represent mean ± SE, n = 4 per treatment group.

Lymphocytes and monocytes are often seen at sites of injury and inflammation [51]. Our lab investigated the effect of soy isoflavone supplementation on ovariectomy induced lymphopoiesis in rats. In this animal study [94], we observed that ovariectomy-induced increases in peripheral blood total lymphocyte and monocyte counts were returned to the levels of sham-operated rats after soy isoflavone supplementation (**Figure 10A** and **B**). Forty-eight 12-month-old Sprague-Dawley rats were either sham-operated (sham; 1 group) or OVX (3 groups) and were fed a standard semi-purified diet for 120 days. Thereafter, the OVX groups received one of the three doses of isoflavones: 0 (OVX), 500 (ISO500), or 1000 (ISO1000) mg/kg diet for 100 days. Ovariectomy significantly ($P < 0.05$) increased the total leukocyte, lymphocyte, monocyte, eosinophil, and basophil counts. Isoflavones at 500 and 1000 mg/kg diet returned the total leukocyte counts as well as leukocyte subpopulations to levels comparable to that of sham. These findings indicate that isoflavones are capable of normalizing circulating levels of inflammatory cells that produce many proinflammatory mediators, which may prove effective for the synovial joint.

Our lab also carried out a three-month double-blind randomized clinical trial [95] to investigate the effects of soy supplementation on symptoms associated with knee OA. About 135 free-living individuals (64 men, mean age = 55.8 ± 13.6 years; and 71 women, mean age = 59.3 ± 12.0 years) with knee OA were randomly assigned to receive 40 g of either soy protein or milk protein daily. This study indicated that soy protein regimen containing 88 mg isoflavones improved ($P < 0.05$) knee range of motion and ability to climb several flights of stairs, and reduced ($P < 0.05$) the intensity, frequency, severity of pain, hindrance to activities (**Figure 11A**), and use of pain medications (**Figure 11B**). The improvement in self-described pain parameters

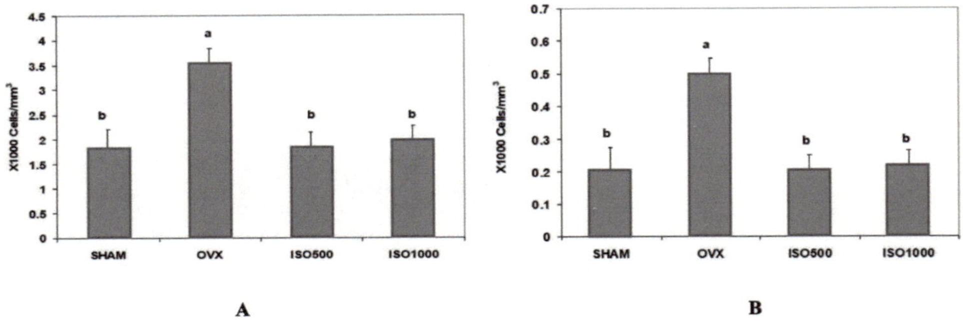

Figure 10.
(A and B) Indicate effects of isoflavones (ISO) on lymphocyte and monocyte counts. Values are mean ± SE (n = 12). Bars that do not share the same superscript are significantly different (P < 0.05).

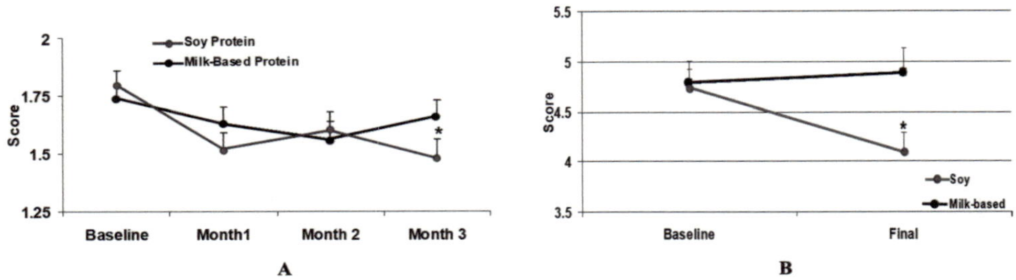

Figure 11.
(A) represents self-reported pain limiting physical activities with scores ranging from 1 to 2; (1) referring to no limitation and (2) referring to pain as causing limitation in physical activity. (B) indicates the use of pain medications (mean + SE). A lower score reflects less use of pain medication and a higher score reflects more frequent use of pain medication.

due to soy supplementation became more pronounced as the treatment duration progressed. Additionally, the soy regimen significantly improved circulating levels of IGF-I which suggests that isoflavones may exert anabolic effects on the cartilage.

In the same study, serum IGF-I as well as human cartilage glycoprotein 39 (YKL-40), a marker of joint destruction [92], were assessed. Results indicated that protein supplementation had significantly lowered mean serum concentration of YKL-40 in men, implying that soy can slow down cartilage degradation. Although both proteins, as expected, increased ($P < 0.05$) circulating levels of IGF-I, soy protein had a more pronounced effect compared to milk protein. We have repeatedly shown [84, 96] that soy has the ability to uniquely enhance serum IGF-I in comparison with milk protein, indicating that this effect may be due to its isoflavone content rather than merely protein.

The findings of our three-month study indicate that soy protein supplementation significantly reduced the intensity and frequency of pain. By comparison, milk protein only reduced pain intensity indicating that the reduction in the frequency of pain and discomfort are specific to soy and not the control protein. Our findings also indicate a reduced need for pain medication. The increased serum IGF-I level with soy supplementation suggests that isoflavones may exert anabolic effects on the cartilage, and decreased YKL-40 levels which is associated with cartilage degeneration, support our hypothesis that soy can improve symptoms and severity of OA. The authors suggest that people with no contraindications to soy isoflavones use ipriflavone, a synthetic isoflavone, for decreasing the symptoms of OA. However, this is just a suggestion and further research must be done to assess the potency of isoflavone usage for symptomatic control of OA.

7. Soy and cardiovascular disease

As mentioned previously, soy isoflavones are phytoestrogens. Estrogen is known to be cardioprotective, so it stands to reason that soy may also be cardioprotective. Many of the clinical trials investigating the effect of soy supplementation on heart health focus mainly on cholesterol levels. This may be due to the fact that the phytosterols, like those found in soy, compete with cholesterol for intestinal absorption [97]. A 2015 study [98] investigated the effect of 8 weeks of standard soymilk supplementation against the effect of 2 g/day of phytosterols and 10 g/day of inulin-enriched soymilk supplementation. While both groups did see a reduction in LDL-C in both groups, the study group supplementing with the extra phytosterols and inulin saw significantly better results. TC was also significantly reduced in the study group, compared to the control of regular soymilk.

Soy can be beneficial in many forms beyond that of soymilk. A study [99] that supplemented whole soy foods (3–4 servings per day) for 12 weeks found that the soy intervention significantly reduced total cholesterol, LDL-C, non-HDL cholesterol, and apoB even though BMI did not decrease. An earlier study [100] also found that soy protein supplementation resulted in decreased cholesterol levels. Prehypertensive women who supplemented 40 g of soy flour saw decreases in LDL-C and well as high sensitivity C-Reactive Protein (CRP), a marker of inflammation [101]. Interestingly, another study found that 1 month of soy nut supplementation modestly reduced arterial stiffness but did not improve inflammatory biomarkers [102]. Additionally, Lucas et al. [103] found that soy isoflavones prevented both hyperlipidemia and atherosclerotic lesions in ovariectomized Golden Syrian Hamsters.

While there are still gaps in the research for CVD and soy consumption, research generally points to a positive effect of soy on heart health, irrespective to its effect on cholesterol. Finding that soy significantly decreased the development of atherosclerotic lesion in a hamster model of postmenopausal CVD is particularly important since CVD remains the leading cause of death in the US.

8. Soy and osteoporosis

Just as OA greatly affects women more so than men, osteoporosis is a particularly concerning problem for the aging female population. Because intestinal cells contain ER, and because estrogen enhances calcium transport [104], it stands to reason that phytoestrogens like soy may increase intestinal calcium transport. There have been multiple studies researching intestinal transport of calcium and soy, as well as the effect of soy on animal models of osteoporosis, and human studies. A study by our lab in 2001 [104] confirmed that not only does ovariectomy decrease rates of calcium transport, but that soy isoflavones in soy protein promoted calcium absorption in a manner analogous to estrogen without any of the side effects/risk. Pawlowski et al. [105] also found that soy isoflavones were effective in increasing calcium retention in bone, and Arjmandi et al. [84] found that women not on hormone replacement therapy who supplemented soy protein experienced reduced urinary calcium excretion.

Animal studies have yielded positive results for isoflavone's bone sparing properties. A 1998 study [106] by our lab compared casein protein and soy protein in ovariectomized (OVX) rats, and found that soy protein with higher levels of isoflavones spared the femoral bone density decreases brought about by ovariectomy. Our 2006 study [107] found that soy positively affected tibial architectural properties of OVX rats, including trabecular thickness, separation, and number without preserving BMD. Another study by our lab [108] found that soy protein with or without isoflavones did not preserve BMD in a male rat model of osteoporosis, but did positively affect the biomechanical properties of bone including yield and ultimate force which are measures of elasticity and plasticity in bone. Multiple other studies have concluded that any bone sparing effects of soy consumption are likely due to soy isoflavone content, which increases bone formation and improves the architectural properties of bone [109–112].

Interestingly, while animal studies have been promising for moderate prevention of bone loss, a 2-year Thai study [113] found that soy isoflavones did not significantly reduce bone loss. Similarly, a 3-year study [114] that gave postmenopausal women soy isoflavones did not find significant bone sparing effects, except for the femoral neck which was still only modestly affected by supplementation. The same lab then evaluated the safety of soy isoflavone supplementation by evaluating effects on hormones, endometrial thickness, and any adverse events, finding no negative evidence of treatment effect on this population, once again indicating that soy supplementation is safe. Wong et al. [115] found that 120 mg of soy isoflavones did reduce whole body BMD loss, but did not positively affect common female fracture sites. Studies by our lab [96, 116], and others [117], generally find that soy supplementation for the treatment of osteoporosis generally has little to no effect on BMD, but may still positively affect bone metabolism as well as bone quality.

9. Conclusions and future directions

Although the role of soy in CVD, lowering cholesterol, and improving bone has been questioned, there is ample evidence to suggest that soy improves symptoms of OA by at least three mechanisms, including (1) acting as a SERM, thereby modulating estrogen receptors; (2) increasing circulating levels of IGF-1, thereby regenerating cartilage and/or preventing further damage; (3) and inhibiting production of inflammatory molecules, such as COX-2, TNF-α. The authors believe that soy plays an important role in the healthy aging process by decreasing the incidence of OA, and allowing those who are afflicted to achieve greater mobility, thus decreasing their chances of developing other chronic diseases that would have resulted from decreased mobility. Therefore, we suggest that consumption of soy and soy isoflavones is crucial for healthy aging and improved QOL throughout the aging process.

We have demonstrated that both leptin and estrogen have a significant role in the etiology, progression, and treatment of OA, but the specifics of that role remain uncertain. The above studies also indicate a positive effect of soy supplementation on cartilage metabolism, inflammation, and indices of pain, likely through the modulation of the aforementioned factors. Soy appears to be promising in the treatment of OA symptoms, but its ability to prevent the disease is questionable. While isoflavones are known to act as SERMs, it is reasonable to suspect that the protein content of soy as a whole in conjunction with isoflavone content is responsible for positive effects in this population. Though the literature indicates that soy supplementation may be helpful in decreasing usage of NSAIDs, slow cartilage degradation, and increase functionality in individuals afflicted with OA, determining the safety as well as the efficacy of soy or its isoflavones as a long-term OA intervention is the next logical step. Any intervention that can improve the QOL of individuals afflicted with OA is worth pursuing, but it is paramount that researchers uncover the exact etiology of the disease so as to prevent future occurrences.

The literature referenced here also indicates that soy can be promising for other chronic disease states, without necessarily posing a risk for increased instance of breast cancer. However, there is still much confusion about which populations are at higher risk for breast cancer when consuming soy. The multiple health benefits appear to outweigh breast cancer risk for most women, even decreasing the chance of breast cancer, but further interventional, rather than strictly epidemiological and cell culture studies, need to be established.

Acknowledgments

We want to thank Dr. Shirin Hooshmand and Jenna Schmidt for collecting and analyzing part of the data.

Conflict of interest

The authors have no conflict of interest to declare.

Author details

Bahram Herman Arjmandi[1,2*] and Elizabeth Marie Foley[1,2]

1 Department of Nutrition, Food and Exercise Sciences, College of Human Sciences, Florida State University, Tallahassee, FL, United States

2 Center for Advancing Exercise and Nutrition Research on Aging (CAENRA), College of Human Sciences, Florida State University, Tallahassee, FL, United States

*Address all correspondence to: barjmandi@fsu.edu

References

[1] Collaborators GM. Global, regional, and national age-sex-specific mortality and life expectancy, 1950-2017: A systematic analysis for the Global Burden of Disease Study 2017. Lancet. 2018;**392**(10159):1684-1735. DOI: 10.1016/S0140-6736(18)31891-9. Epub 2018/11/08. PubMed PMID: 30496102; PubMed Central PMCID: PMCPMC6227504

[2] Xu J, Murphy SL, Kochanek KD, Arias E. Mortality in the United States, 2015. NCHS Data Briefs. 2016;**267**:1-8. PubMed PMID: 27930283

[3] Brown GC. Living too long: The current focus of medical research on increasing the quantity, rather than the quality, of life is damaging our health and harming the economy. EMBO Reports. 2015;**16**(2):137-141. DOI: 10.15252/embr.201439518. Epub 2014/12/18. PubMed PMID: 25525070; PubMed Central PMCID: PMCPMC4328740

[4] Shlisky J, Bloom DE, Beaudreault AR, Tucker KL, Keller HH, Freund-Levi Y, et al. Nutritional considerations for healthy aging and reduction in age-related chronic disease. Advances in Nutrition. 2017;**8**(1):17-26. DOI: 10.3945/an.116.013474. Epub 2017/01/17. PubMed PMID: 28096124; PubMed Central PMCID: PMCPMC5227979

[5] Nunan D, Mahtani KR, Roberts N, Heneghan C. Physical activity for the prevention and treatment of major chronic disease: An overview of systematic reviews. Systematic Reviews. 2013;**2**:56. DOI: 10.1186/2046-4053-2-56. Epub 2013/07/10. PubMed PMID: 23837523; PubMed Central PMCID: PMCPMC3710239

[6] Cross M, Smith E, Hoy D, Nolte S, Ackerman I, Fransen M, et al. The global burden of hip and knee osteoarthritis: Estimates from the global burden of disease 2010 study. Annals of the Rheumatic Diseases. 2014;**73**(7):1323-1330. DOI: 10.1136/annrheumdis-2013-204763. Epub 2014/02/19. PubMed PMID: 24553908

[7] Day JS, Van Der Linden JC, Bank RA, Ding M, Hvid I, Sumner DR, et al. Adaptation of subchondral bone in osteoarthritis. Biorheology. 2004;**41**(3-4):359-368. PubMed PMID: 15299268

[8] Felson DT, Naimark A, Anderson J, Kazis L, Castelli W, Meenan RF. The prevalence of knee osteoarthritis in the elderly. The Framingham osteoarthritis study. Arthritis and Rheumatism. 1987;**30**(8):914-918. PubMed PMID: 3632732

[9] Pereira D, Peleteiro B, Araújo J, Branco J, Santos RA, Ramos E. The effect of osteoarthritis definition on prevalence and incidence estimates: A systematic review. Osteoarthritis and Cartilage. 2011;**19**(11):1270-1285. DOI: 10.1016/j.joca.2011.08.009. Epub 2011/08/24. PubMed PMID: 21907813

[10] Heidari B. Knee osteoarthritis prevalence, risk factors, pathogenesis and features: Part I. Caspian Journal of Internal Medicine. 2011;**2**(2):205-212. PubMed PMID: 24024017; PubMed Central PMCID: PMCPMC3766936

[11] Paul RF, Hassan M, Nazar HS, Gillani S, Afzal N, Qayyum I. Effect of body mass index on serum leptin levels. Journal of Ayub Medical College, Abbottabad. 2011;**23**(3):40-43. PubMed PMID: 23272432

[12] Al Maskari MY, Alnaqdy AA. Correlation between serum leptin levels, body mass index and obesity in Omanis. Sultan Qaboos University Medical Journal. 2006;**6**(2):27-31. PubMed PMID: 21748132; PubMed Central PMCID: PMCPMC3074914

[13] Eymard F, Parsons C, Edwards MH, Petit-Dop F, Reginster JY, Bruyère O, et al. Diabetes is a risk factor for knee osteoarthritis progression. Osteoarthritis and Cartilage. 2015;**23**(6):851-859. DOI: 10.1016/j.joca.2015.01.013. Epub 2015/02/03. PubMed PMID: 25655678

[14] Minihane AM, Vinoy S, Russell WR, Baka A, Roche HM, Tuohy KM, et al. Low-grade inflammation, diet composition and health: Current research evidence and its translation. The British Journal of Nutrition. 2015;**114**(7):999-1012. DOI: 10.1017/S0007114515002093. Epub 2015/07/31. PubMed PMID: 26228057; PubMed Central PMCID: PMCPMC4579563

[15] Brooks P. Inflammation as an important feature of osteoarthritis. Bulletin of the World Health Organization. 2003;**81**(9):689-690. Epub 2003/11/14. PubMed PMID: 14710513; PubMed Central PMCID: PMCPMC2572543

[16] Poole AR, Nelson F, Dahlberg L, Tchetina E, Kobayashi M, Yasuda T, et al. Proteolysis of the collagen fibril in osteoarthritis. Biochemical Society Symposium. 2003;**70**:115-123. PubMed PMID: 14587287

[17] Brenner SS, Klotz U, Alscher DM, Mais A, Lauer G, Schweer H, et al. Osteoarthritis of the knee—Clinical assessments and inflammatory markers. Osteoarthritis and Cartilage. 2004;**12**(6):469-475. DOI: 10.1016/j.joca.2004.02.011. PubMed PMID: 15135143

[18] Ishiguro N, Kojima T, Poole AR. Mechanism of cartilage destruction in osteoarthritis. Nagoya Journal of Medical Science. 2002;**65**(3-4):73-84. PubMed PMID: 12580533

[19] Nakamura H, Shibakawa A, Tanaka M, Kato T, Nishioka K. Effects of glucosamine hydrochloride on the production of prostaglandin E2, nitric oxide and metalloproteases by chondrocytes and synoviocytes in osteoarthritis. Clinical and Experimental Rheumatology. 2004;**22**(3):293-299. PubMed PMID: 15144122

[20] Anderson JW, Johnstone BM, Cook-Newell ME. Meta-analysis of the effects of soy protein intake on serum lipids. The New England Journal of Medicine. 1995;**333**(5):276-282. DOI: 10.1056/NEJM199508033330502. PubMed PMID: 7596371

[21] Lu S, Nishimura K, Hossain MA, Jisaka M, Nagaya T, Yokota K. Regulation and role of arachidonate cascade during changes in life cycle of adipocytes. Applied Biochemistry and Biotechnology. 2004;**118**(1-3):133-153. PubMed PMID: 15304745

[22] Tamura M, Deb S, Sebastian S, Okamura K, Bulun SE. Estrogen up-regulates cyclooxygenase-2 via estrogen receptor in human uterine microvascular endothelial cells. Fertility and Sterility. 2004;**81**(5):1351-1356. DOI: 10.1016/j.fertnstert.2003.09.076. PubMed PMID: 15136101

[23] Montgomery KS. Soy protein. The Journal of Perinatal Education. 2003;**12**(3):42-45. DOI: 10.1624/105812403X106946. PubMed PMID: 17273351; PubMed Central PMCID: PMCPMC1595159

[24] Hughes GJ, Ryan DJ, Mukherjea R, Schasteen CS. Protein digestibility-corrected amino acid scores (PDCAAS) for soy protein isolates and concentrate: Criteria for evaluation. Journal of Agricultural and Food Chemistry. 2011;**59**(23):12707-12712. DOI: 10.1021/jf203220v. Epub 2011/11/16. PubMed PMID: 22017752

[25] Bang MH, Chio OS, Kim WK. Soyoligosaccharide increases fecal bifidobacteria counts, short-chain fatty acids, and fecal lipid concentrations in young Korean women. Journal of Medicinal Food. 2007;**10**(2):366-370.

DOI: 10.1089/jmf.2005.096. PubMed PMID: 17651076

[26] Slavin M, Kenworthy W, Yu LL. Antioxidant properties, phytochemical composition, and antiproliferative activity of Maryland-grown soybeans with colored seed coats. Journal of Agricultural and Food Chemistry. 2009;**57**(23):11174-11185. DOI: 10.1021/jf902609n. PubMed PMID: 19950996

[27] Messina M. Soy and health update: Evaluation of the clinical and epidemiologic literature. Nutrients. 2016;**8**(12):1-42. DOI: 10.3390/nu8120754. Epub 2016/11/24. PubMed PMID: 27886135; PubMed Central PMCID: PMCPMC5188409

[28] Setchell KD, Brown NM, Zimmer-Nechemias L, Brashear WT, Wolfe BE, Kirschner AS, et al. Evidence for lack of absorption of soy isoflavone glycosides in humans, supporting the crucial role of intestinal metabolism for bioavailability. The American Journal of Clinical Nutrition. 2002;**76**(2):447-453. DOI: 10.1093/ajcn/76.2.447. PubMed PMID: 12145021

[29] Nakajima N, Nozaki N, Ishihara K, Ishikawa A, Tsuji H. Analysis of isoflavone content in tempeh, a fermented soybean, and preparation of a new isoflavone-enriched tempeh. Journal of Bioscience and Bioengineering. 2005;**100**(6):685-687. DOI: 10.1263/jbb.100.685. PubMed PMID: 16473782

[30] Setchell KDR. The history and basic science development of soy isoflavones. Menopause. 2017;**24**(12):1338-1350. DOI: 10.1097/GME.0000000000001018. PubMed PMID: 29189602

[31] Charalambous C, Pitta CA, Constantinou AI. Equol enhances tamoxifen's anti-tumor activity by induction of caspase-mediated apoptosis in MCF-7 breast cancer cells. BMC Cancer. 2013;**13**:238. DOI: 10.1186/1471-2407-13-238. Epub 2013/05/15. PubMed PMID: 23675643; PubMed Central PMCID: PMCPMC3661348

[32] Lecomte S, Demay F, Ferrière F, Pakdel F. Phytochemicals targeting estrogen receptors: Beneficial rather than adverse effects? International Journal of Molecular Sciences. 2017;**18**(7):1-19. DOI: 10.3390/ijms18071381. Epub 2017/06/28. PubMed PMID: 28657580; PubMed Central PMCID: PMCPMC5535874

[33] DeSantis C, Ma J, Bryan L, Jemal A. Breast cancer statistics, 2013. CA: A Cancer Journal for Clinicians. 2014;**64**(1):52-62. DOI: 10.3322/caac.21203. Epub 2013/10/01. PubMed PMID: 24114568

[34] Iversen A, Thune I, McTiernan A, Emaus A, Finstad SE, Flote V, et al. Ovarian hormones and reproductive risk factors for breast cancer in premenopausal women: The Norwegian EBBA-I study. Human Reproduction. 2011;**26**(6):1519-1529. DOI: 10.1093/humrep/der081. Epub 2011/04/05. PubMed PMID: 21467202; PubMed Central PMCID: PMCPMC3096559

[35] Samavat H, Kurzer MS. Estrogen metabolism and breast cancer. Cancer Letters. 2015;**356**(2 Pt A):231-243. DOI: 10.1016/j.canlet.2014.04.018. Epub 2014/04/28. PubMed PMID: 24784887; PubMed Central PMCID: PMCPMC4505810

[36] Roodi N, Bailey LR, Kao WY, Verrier CS, Yee CJ, Dupont WD, et al. Estrogen receptor gene analysis in estrogen receptor-positive and receptor-negative primary breast cancer. Journal of the National Cancer Institute. 1995;**87**(6): 446-451. PubMed PMID: 7861463

[37] Douglas CC, Johnson SA, Arjmandi BH. Soy and its isoflavones: The truth behind the science in breast cancer. Anti-Cancer Agents in Medicinal

Chemistry. 2013;**13**(8):1178-1187. PubMed PMID: 23919747

[38] Anderson JJ, Anthony MS, Cline JM, Washburn SA, Garner SC. Health potential of soy isoflavones for menopausal women. Public Health Nutrition. 1999;**2**(4):489-504. PubMed PMID: 10656468

[39] Kulkarni KH, Yang Z, Niu T, Hu M. Effects of estrogen and estrus cycle on pharmacokinetics, absorption, and disposition of genistein in female Sprague-Dawley rats. The Journal of Agricultural and Food Chemistry. 2012;**60**(32):7949-7956. DOI: 10.1021/jf204755g. Epub 2012/08/03. PubMed PMID: 22757747; PubMed Central PMCID: PMCPMC4030716

[40] Wu AH, Yu MC, Tseng CC, Pike MC. Epidemiology of soy exposures and breast cancer risk. The British Journal of Cancer. 2008;**98**(1):9-14. DOI: 10.1038/sj.bjc.6604145. Epub 2008/01/08. PubMed PMID: 18182974; PubMed Central PMCID: PMCPMC2359677

[41] Verheus M, van Gils CH, Keinan-Boker L, Grace PB, Bingham SA, Peeters PH. Plasma phytoestrogens and subsequent breast cancer risk. Journal of Clinical Oncology. 2007;**25**(6):648-655. DOI: 10.1200/JCO.2006.06.0244. Epub 2007/01/02. PubMed PMID: 17200150

[42] Shu XO, Zheng Y, Cai H, Gu K, Chen Z, Zheng W, et al. Soy food intake and breast cancer survival. Journal of the American Medical Association. 2009;**302**(22):2437-2443. DOI: 10.1001/jama.2009.1783. PubMed PMID: 19996398; PubMed Central PMCID: PMCPMC2874068

[43] Zava DT, Duwe G. Estrogenic and antiproliferative properties of genistein and other flavonoids in human breast cancer cells in vitro. Nutrition and Cancer. 1997;**27**(1):31-40. DOI: 10.1080/01635589709514498. PubMed PMID: 8970179

[44] Mizushina Y, Shiomi K, Kuriyama I, Takahashi Y, Yoshida H. Inhibitory effects of a major soy isoflavone, genistein, on human DNA topoisomerase II activity and cancer cell proliferation. International Journal of Oncology. 2013;**43**(4):1117-1124. DOI: 10.3892/ijo.2013.2032. Epub 2013/07/23. PubMed PMID: 23900272

[45] Davis TA, Mungunsukh O, Zins S, Day RM, Landauer MR. Genistein induces radioprotection by hematopoietic stem cell quiescence. International Journal of Radiation Biology. 2008;**84**(9):713-726. DOI: 10.1080/09553000802317778. PubMed PMID: 18821385

[46] Magee PJ, McGlynn H, Rowland IR. Differential effects of isoflavones and lignans on invasiveness of MDA-MB-231 breast cancer cells in vitro. Cancer Letters. 2004;**208**(1):35-41. DOI: 10.1016/j.canlet.2003.11.012. PubMed PMID: 15105043

[47] Shike M, Doane AS, Russo L, Cabal R, Reis-Filho JS, Gerald W, et al. The effects of soy supplementation on gene expression in breast cancer: A randomized placebo-controlled study. Journal of the National Cancer Institute. 2014;**106**(9):1-12. DOI: 10.1093/jnci/dju189. Epub 2014/09/04. PubMed PMID: 25190728; PubMed Central PMCID: PMCPMC4817128

[48] Alexandersen P, Toussaint A, Christiansen C, Devogelaer JP, Roux C, Fechtenbaum J, et al. Ipriflavone in the treatment of postmenopausal osteoporosis: A randomized controlled trial. Journal of the American Medical Association. 2001;**285**(11):1482-1488. PubMed PMID: 11255425

[49] Agnusdei D, Bufalino L. Efficacy of ipriflavone in established osteoporosis and long-term safety. Calcified Tissue International. 1997;**61**(Suppl 1):S23-S27. PubMed PMID: 9263613

[50] Ben-Hur H, Mor G, Insler V, Blickstein I, Amir-Zaltsman Y, Sharp A, et al. Menopause is associated with a significant increase in blood monocyte number and a relative decrease in the expression of estrogen receptors in human peripheral monocytes. American Journal of Reproductive Immunology. 1995;**34**(6):363-369. PubMed PMID: 8607941

[51] Soung DY, Khalil DA, Arquitt AB, Smith BJ, Hammond LJ, Droke EA, et al. Soy isoflavones prevent the ovarian hormone deficiency-associated rise in leukocytes in rats. Phytomedicine. 2004;**11**(4):303-308. DOI: 10.1078/0944711041495164. PubMed PMID: 15185842

[52] Soung DY, Patade A, Khalil DA, Lucas EA, Devareddy L, Greaves KA, et al. Soy protein supplementation does not cause lymphocytopenia in postmenopausal women. Nutrition Journal. 2006;**5**:12. DOI: 10.1186/1475-2891-5-12. Epub 2006/04/11. PubMed PMID: 16608514; PubMed Central PMCID: PMCPMC1481570

[53] Martín-Millán M, Castañeda S. Estrogens, osteoarthritis and inflammation. Joint Bone Spine. 2013;**80**(4):368-373. DOI: 10.1016/j.jbspin.2012.11.008. Epub 2013/01/23. PubMed PMID: 23352515

[54] Inoue K, Ushiyama T, Kim Y, Shichikawa K, Nishioka J, Hukuda S. Increased rate of hysterectomy in women undergoing surgery for osteoarthritis of the knee. Osteoarthritis Cartilage. 1995;**3**(3):205-209. PubMed PMID: 8581750

[55] Spector TD, Brown GC, Silman AJ. Increased rates of previous hysterectomy and gynaecological operations in women with osteoarthritis. BMJ. 1988;**297**(6653):899-900. PubMed PMID: 3140970; PubMed Central PMCID: PMCPMC1834435

[56] Spector TD, Hart DJ, Brown P, Almeyda J, Dacre JE, Doyle DV, et al. Frequency of osteoarthritis in hysterectomized women. Journal of Rheumatology. 1991;**18**(12):1877-1883. PubMed PMID: 1795326

[57] Richette P, Corvol M, Bardin T. Estrogens, cartilage, and osteoarthritis. Joint Bone Spine. 2003;**70**(4):257-262. PubMed PMID: 12951307

[58] Stöve J, Stürmer T, Kessler S, Brenner H, Puhl W, Günther KP. Hysterectomy and patterns of osteoarthritis. The Ulm Osteoarthritis Study. Scandinavian Journal of Rheumatology. 2001;**30**(6):340-345. PubMed PMID: 11846052

[59] Gao W, Zeng C, Cai D, Liu B, Li Y, Wen X, et al. Serum concentrations of selected endogenous estrogen and estrogen metabolites in pre- and post-menopausal Chinese women with osteoarthritis. The Journal of Endocrinological Investigation. 2010;**33**(9):644-649. DOI: 10.1007/BF03346664. Epub 2010/03/25. PubMed PMID: 20339312

[60] Roman-Blas JA, Castañeda S, Largo R, Herrero-Beaumont G. Osteoarthritis associated with estrogen deficiency. Arthritis Research and Therapy. 2009;**11**(5):241. DOI: 10.1186/ar2791. Epub 2009/09/21. PubMed PMID: 19804619; PubMed Central PMCID: PMCPMC2787275

[61] Tsai CL, Liu TK, Chen TJ. Estrogen and osteoarthritis: A study of synovial estradiol and estradiol receptor binding in human osteoarthritic knees. Biochemical and Biophysical Research Communications. 1992;**183**(3): 1287-1291. PubMed PMID: 1567405

[62] Rosner IA, Malemud CJ, Goldberg VM, Papay RS, Getzy L, Moskowitz RW. Pathologic and metabolic responses of experimental osteoarthritis to estradiol and

an estradiol antagonist. Clinical Orthopaedics and Related Research. 1982;(171):280-286. PubMed PMID: 7140079

[63] Ng MC, Harper RP, Le CT, Wong BS. Effects of estrogen on the condylar cartilage of the rat mandible in organ culture. Journal of Oral and Maxillofacial Surgery. 1999;**57**(7): 818-823. PubMed PMID: 10416629

[64] Tsai CL, Liu TK. Up-regulation of estrogen receptors in rabbit osteoarthritic cartilage. Life Sciences. 1992;**50**(22):1727-1735. PubMed PMID: 1588803

[65] Setchell KD. Soy isoflavones—Benefits and risks from nature's selective estrogen receptor modulators (SERMs). Journal of the American College of Nutrition. 2001;**20**(5 Suppl):354S-362S; discussion 81S–83S. PubMed PMID: 11603644

[66] Zhang Z, Li L, Yang W, Cao Y, Shi Y, Li X, et al. The effects of different doses of IGF-1 on cartilage and subchondral bone during the repair of full-thickness articular cartilage defects in rabbits. Osteoarthritis Cartilage. 2017;**25**(2):309-320. DOI: 10.1016/j.joca.2016.09.010. Epub 2016/09/20. PubMed PMID: 27662821

[67] Foley E, Browne J, Akhavan N, George K, Muños J, Siebert S, et al. Relationship between inflammation, oxidative damage, weight, and severity of knee osteoarthritis. ASN 2018. Abstract. 2018

[68] Akhavan NS, Ormsbee L, Johnson SA, George KS, Foley EM, Elam ML, et al. Functionality in middle-aged and older overweight and obese individuals with knee osteoarthritis. Healthcare (Basel). 2018;**6**(3):1-12. DOI: 10.3390/healthcare6030074. Epub 2018/07/04. PubMed PMID: 29973574

[69] Ahima RS. Revisiting leptin's role in obesity and weight loss. The Journal of Clinical Investigation. 2008;**118**(7):2380-2383. DOI: 10.1172/JCI36284. PubMed PMID: 18568083; PubMed Central PMCID: PMCPMC2430504

[70] Scotece M, Mobasheri A. Leptin in osteoarthritis: Focus on articular cartilage and chondrocytes. Life Sciences. 2015;**140**:75-78. DOI: 10.1016/j.lfs.2015.05.025. Epub 2015/06/19. PubMed PMID: 26094910

[71] Farr OM, Gavrieli A, Mantzoros CS. Leptin applications in 2015: What have we learned about leptin and obesity? Current Opinion in Endocrinology Diabetes and Obesity. 2015;**22**(5):353-359. DOI: 10.1097/MED.0000000000000184. PubMed PMID: 26313897; PubMed Central PMCID: PMCPMC4610373

[72] Vuolteenaho K, Koskinen A, Moilanen E. Leptin—A link between obesity and osteoarthritis: Applications for prevention and treatment. Basic and Clinical Pharmacology and Toxicology. 2014;**114**(1):103-108. DOI: 10.1111/bcpt.12160. Epub 2013/11/20. PubMed PMID: 24138453

[73] Rosenbaum M, Nicolson M, Hirsch J, Heymsfield SB, Gallagher D, Chu F, et al. Effects of gender, body composition, and menopause on plasma concentrations of leptin. Journal of Clinical Endocrinology and Metabolism. 1996;**81**(9):3424-3427. DOI: 10.1210/jcem.81.9.8784109. PubMed PMID: 8784109

[74] Bao JP, Chen WP, Feng J, Hu PF, Shi ZL, Wu LD. Leptin plays a catabolic role on articular cartilage. Molecular Biology Reports. 2010;**37**(7):3265-3272. DOI: 10.1007/s11033-009-9911-x. Epub 2009/10/30. PubMed PMID: 19876764

[75] Presle N, Pottie P, Dumond H, Guillaume C, Lapicque F, Pallu S, et al. Differential distribution of adipokines between serum and

synovial fluid in patients with osteoarthritis. Contribution of joint tissues to their articular production. Osteoarthritis Cartilage. 2006;**14**(7):690-695. DOI: 10.1016/j.joca.2006.01.009. Epub 2006/03/09. PubMed PMID: 16527497

[76] Schmidt J, Shirin H, Arjmandi B. Relationship between serum and synovial fluid concentration of leptin and degree of osteoarthritis. FASEB Journal. Abstract. 2011

[77] La Cava A, Matarese G. The weight of leptin in immunity. Nature Reviews Immunology. 2004;**4**(5):371-379. DOI: 10.1038/nri1350. PubMed PMID: 15122202

[78] Siegmund B, Lehr HA, Fantuzzi G. Leptin: A pivotal mediator of intestinal inflammation in mice. Gastroenterology. 2002;**122**(7):2011-2025. PubMed PMID: 12055606

[79] Iikuni N, Lam QL, Lu L, Matarese G, La Cava A. Leptin and Inflammation. Current Immunology Reviews. 2008;**4**(2):70-79. DOI: 10.2174/157339508784325046. PubMed PMID: 20198122; PubMed Central PMCID: PMCPMC2829991

[80] Frigolet ME, Torres N, Uribe-Figueroa L, Rangel C, Jimenez-Sanchez G, Tovar AR. White adipose tissue genome wide-expression profiling and adipocyte metabolic functions after soy protein consumption in rats. Journal of Nutritional Biochemistry. 2011;**22**(2):118-129. DOI: 10.1016/j.jnutbio.2009.12.006. Epub 2010/05/14. PubMed PMID: 20471815

[81] Niwa T, Yokoyama S, Osawa T. Effect of the genistein metabolite on leptin secretion in murine adipocytes in vitro. Food Chemistry. 2013;**138**(1):122-125. DOI: 10.1016/j.foodchem.2012.09.108. Epub 2012/11/08. PubMed PMID: 23265465

[82] Llaneza P, González C, Fernández-Iñarrea J, Alonso A, Díaz F, Pérez-López FR. Soy isoflavones improve insulin sensitivity without changing serum leptin among postmenopausal women. Climacteric. 2012;**15**(6):611-620. DOI: 10.3109/13697137.2011.631062. Epub 2011/12/23. PubMed PMID: 22191384

[83] Kwak JH, Ahn CW, Park SH, Jung SU, Min BJ, Kim OY, et al. Weight reduction effects of a black soy peptide supplement in overweight and obese subjects: Double blind, randomized, controlled study. Food and Function. 2012;**3**(10):1019-1024. DOI: 10.1039/c2fo10244g. Epub 2012/06/28. PubMed PMID: 22739624

[84] Arjmandi BH, Khalil DA, Smith BJ, Lucas EA, Juma S, Payton ME, et al. Soy protein has a greater effect on bone in postmenopausal women not on hormone replacement therapy, as evidenced by reducing bone resorption and urinary calcium excretion. Journal of Clinical Endocrinology and Metabolism. 2003;**88**(3):1048-1054. DOI: 10.1210/jc.2002-020849. PubMed PMID: 12629084

[85] Ye F, Wu J, Dunn T, Yi J, Tong X, Zhang D. Inhibition of cyclooxygenase-2 activity in head and neck cancer cells by genistein. Cancer Letters. 2004;**211**(1):39-46. DOI: 10.1016/j.canlet.2004.03.043. PubMed PMID: 15194215

[86] Liang YC, Huang YT, Tsai SH, Lin-Shiau SY, Chen CF, Lin JK. Suppression of inducible cyclooxygenase and inducible nitric oxide synthase by apigenin and related flavonoids in mouse macrophages. Carcinogenesis. 1999;**20**(10):1945-1952. PubMed PMID: 10506109

[87] Hooshmand S, Soung DY, Lucas EA, Madihally SV, Levenson CW, Arjmandi BH. Genistein reduces the production of proinflammatory molecules in human chondrocytes. Journal of Nutritional

Biochemistry. 2007;**18**(9):609-614. DOI: 10.1016/j.jnutbio.2006.11.006. Epub 2007/03/21. PubMed PMID: 17368882

[88] Zidar N, Odar K, Glavac D, Jerse M, Zupanc T, Stajer D. Cyclooxygenase in normal human tissues—Is COX-1 really a constitutive isoform, and COX-2 an inducible isoform? Journal of Cellular and Molecular Medicine. 2009;**13**(9B): 3753-3763. DOI: 10.1111/j.1582-4934. 2008.00430.x. Epub 2008/07/24. PubMed PMID: 18657230; PubMed Central PMCID: PMCPMC4516524

[89] Dubois RW, Melmed GY, Henning JM, Bernal M. Risk of upper gastrointestinal injury and events in patients treated with cyclooxygenase (COX)-1/COX-2 nonsteroidal antiinflammatory drugs (NSAIDs), COX-2 selective NSAIDs, and gastroprotective cotherapy: An appraisal of the literature. Journal of Clinical Rheumatology. 2004;**10**(4):178-189. DOI: 10.1097/01. rhu.0000128851.12010.46. PubMed PMID: 17043507

[90] Hörl WH. Nonsteroidal anti-inflammatory drugs and the kidney. Pharmaceuticals (Basel). 2010;**3**(7):2291-2321. DOI: 10.3390/ph3072291. Epub 2010/07/21. PubMed PMID: 27713354; PubMed Central PMCID: PMCPMC4036662

[91] Hawkey CJ. COX-1 and COX-2 inhibitors. Best Practice and Research Clinical Gastroenterology. 2001;**15**(5): 801-820. DOI: 10.1053/bega.2001.0236. PubMed PMID: 11566042

[92] Volck B, Johansen JS, Stoltenberg M, Garbarsch C, Price PA, Ostergaard M, et al. Studies on YKL-40 in knee joints of patients with rheumatoid arthritis and osteoarthritis. Involvement of YKL-40 in the joint pathology. Osteoarthritis Cartilage. 2001;**9**(3):203-214. DOI: 10.1053/joca.2000.0377. PubMed PMID: 11300743

[93] Borzan J, Tall JM, Zhao C, Meyer RA, Raja SN. Effects of soy diet on inflammation-induced primary and secondary hyperalgesia in rat. The European Journal of Pain. 2010;**14**(8):792-798. DOI: 10.1016/j. ejpain.2009.12.002. Epub 2010/01/08. PubMed PMID: 20060762; PubMed Central PMCID: PMCPMC2891824

[94] Hurtubise J, McLellan K, Durr K, Onasanya O, Nwabuko D, Ndisang JF. The different facets of dyslipidemia and hypertension in atherosclerosis. Current Atherosclerosis Reports. 2016;**18**(12):82. DOI: 10.1007/s11883-016-0632-z. PubMed PMID: 27822682

[95] Arjmandi BH, Khalil DA, Lucas EA, Smith BJ, Sinichi N, Hodges SB, et al. Soy protein may alleviate osteoarthritis symptoms. Phytomedicine. 2004;**11**(7-8):567-575. DOI: 10.1016/j. phymed.2003.11.001. PubMed PMID: 15636169

[96] Khalil DA, Lucas EA, Juma S, Smith BJ, Payton ME, Arjmandi BH. Soy protein supplementation increases serum insulin-like growth factor-I in young and old men but does not affect markers of bone metabolism. Journal of Nutrition. 2002;**132**(9):2605-2608. DOI: 10.1093/jn/132.9.2605. PubMed PMID: 12221217

[97] Katan MB, Grundy SM, Jones P, Law M, Miettinen T, Paoletti R, et al. Efficacy and safety of plant stanols and sterols in the management of blood cholesterol levels. Mayo Clinic Proceedings. 2003;**78**(8):965-978. DOI: 10.4065/78.8.965. PubMed PMID: 12911045

[98] Kietsiriroje N, Kwankaew J, Kitpakornsanti S, Leelawattana R. Effect of phytosterols and inulin-enriched soymilk on LDL-cholesterol in Thai subjects: A double-blinded randomized controlled trial. Lipids in Health and Disease. 2015;**14**:146. DOI: 10.1186/s12944-015-0149-4. Epub 2015/11/09.

PubMed PMID: 26553006; PubMed Central PMCID: PMCPMC4640379

[99] Ruscica M, Pavanello C, Gandini S, Gomaraschi M, Vitali C, Macchi C, et al. Effect of soy on metabolic syndrome and cardiovascular risk factors: A randomized controlled trial. The European Journal of Nutrition. 2018;**57**(2):499-511. DOI: 10.1007/s00394-016-1333-7. Epub 2016/10/18. PubMed PMID: 27757595

[100] Harland JI, Haffner TA. Systematic review, meta-analysis and regression of randomised controlled trials reporting an association between an intake of circa 25 g soya protein per day and blood cholesterol. Atherosclerosis. 2008;**200**(1):13-27. DOI: 10.1016/j.atherosclerosis.2008.04.006. Epub 2008/04/15. PubMed PMID: 18534601

[101] Liu ZM, Ho SC, Chen YM, Ho S, To K, Tomlinson B, et al. Whole soy, but not purified daidzein, had a favorable effect on improvement of cardiovascular risks: A 6-month randomized, double-blind, and placebo-controlled trial in equol-producing postmenopausal women. Molecular Nutrition and Food Research. 2014;**58**(4):709-717. DOI: 10.1002/mnfr.201300499. Epub 2013/11/24. PubMed PMID: 24273218

[102] Reverri EJ, LaSalle CD, Franke AA, Steinberg FM. Soy provides modest benefits on endothelial function without affecting inflammatory biomarkers in adults at cardiometabolic risk. Molecular Nutrition and Food Research. 2015;**59**(2):323-333. DOI: 10.1002/mnfr.201400270. Epub 2014/12/05. PubMed PMID: 25351805; PubMed Central PMCID: PMCPMC4451218

[103] Lucas EA, Lightfoot SA, Hammond LJ, Devareddy L, Khalil DA, Daggy BP, et al. Soy isoflavones prevent ovariectomy-induced atherosclerotic lesions in Golden Syrian hamster model of postmenopausal hyperlipidemia. Menopause. 2003;**10**(4):314-321. DOI: 10.1097/01.GME.0000051509.84118.FD. PubMed PMID: 12851514

[104] Arjmandi BH, Khalil DA, Hollis BW. Soy protein: Its effects on intestinal calcium transport, serum vitamin D, and insulin-like growth factor-I in ovariectomized rats. Calcified Tissue International. 2002;**70**(6):483-487. DOI: 10.1007/s00223-001-1100-4. Epub 2002/06/01. PubMed PMID: 27695965

[105] Pawlowski JW, Martin BR, McCabe GP, McCabe L, Jackson GS, Peacock M, et al. Impact of equol-producing capacity and soy-isoflavone profiles of supplements on bone calcium retention in postmenopausal women: A randomized crossover trial. The American Journal of Clinical Nutrition. 2015;**102**(3):695-703. DOI: 10.3945/ajcn.114.093906. Epub 2015/08/05. PubMed PMID: 26245807; PubMed Central PMCID: PMCPMC4548170

[106] Arjmandi BH, Birnbaum R, Goyal NV, Getlinger MJ, Juma S, Alekel L, et al. Bone-sparing effect of soy protein in ovarian hormone-deficient rats is related to its isoflavone content. The American Journal of Clinical Nutrition. 1998;**68**(6 Suppl):1364S-1368S. DOI: 10.1093/ajcn/68.6.1364S. PubMed PMID: 9848500

[107] Devareddy L, Khalil DA, Smith BJ, Lucas EA, Soung DY, Marlow DD, et al. Soy moderately improves microstructural properties without affecting bone mass in an ovariectomized rat model of osteoporosis. Bone. 2006;**38**(5):686-693. DOI: 10.1016/j.bone.2005.10.024. Epub 2006/01/10. PubMed PMID: 16406762

[108] Juma SS, Ezzat-Zadeh Z, Khalil DA, Hooshmand S, Akhter M, Arjmandi BH. Soy protein with or without isoflavones failed to preserve bone density in gonadal hormone-deficient male rat model of osteoporosis. Nutrition Research. 2012;**32**(9):694-700. DOI:

10.1016/j.nutres.2012.08.001. Epub 2012/09/23. PubMed PMID: 23084642

[109] Soung DY, Devareddy L, Khalil DA, Hooshmand S, Patade A, Lucas EA, et al. Soy affects trabecular microarchitecture and favorably alters select bone-specific gene expressions in a male rat model of osteoporosis. Calcified Tissue International. 2006;**78**(6):385-391. DOI: 10.1007/s00223-005-0069-9. Epub 2006/07/21. PubMed PMID: 16830200

[110] Devareddy L, Khalil DA, Korlagunta K, Hooshmand S, Bellmer DD, Arjmandi BH. The effects of fructo-oligosaccharides in combination with soy protein on bone in osteopenic ovariectomized rats. Menopause. 2006;**13**(4):692-699. DOI: 10.1097/01.gme.0000195372.74944.71. PubMed PMID: 16837891

[111] Khalil DA, Lucas EA, Smith BJ, Soung DY, Devareddy L, Juma S, et al. Soy isoflavones may protect against orchidectomy-induced bone loss in aged male rats. Calcified Tissue International. 2005;**76**(1):56-62. DOI: 10.1007/s00223-004-0018-z. Epub 2004/11/04. PubMed PMID: 15549639

[112] Hooshmand S, Juma S, Arjmandi BH. Combination of genistin and fructooligosaccharides prevents bone loss in ovarian hormone deficiency. Journal of Medicinal Food. 2010;**13**(2):320-325. DOI: 10.1089/jmf.2009.0059. PubMed PMID: 20132047

[113] Tai TY, Tsai KS, Tu ST, Wu JS, Chang CI, Chen CL, et al. The effect of soy isoflavone on bone mineral density in postmenopausal Taiwanese women with bone loss: A 2-year randomized double-blind placebo-controlled study. Osteoporosis International. 2012;**23**(5):1571-1580. DOI: 10.1007/s00198-011-1750-7. Epub 2011/09/08. PubMed PMID: 21901480; PubMed Central PMCID: PMCPMC3332377

[114] Alekel DL, Van Loan MD, Koehler KJ, Hanson LN, Stewart JW, Hanson KB, et al. The soy isoflavones for reducing bone loss (SIRBL) study: A 3-y randomized controlled trial in postmenopausal women. The American Journal of Clinical Nutrition. 2010;**91**(1):218-230. DOI: 10.3945/ajcn.2009.28306. Epub 2009/11/11. PubMed PMID: 19906801; PubMed Central PMCID: PMCPMC2793109

[115] Wong WW, Lewis RD, Steinberg FM, Murray MJ, Cramer MA, Amato P, et al. Soy isoflavone supplementation and bone mineral density in menopausal women: A 2-y multicenter clinical trial. The American Journal of Clinical Nutrition. 2009;**90**(5):1433-1439. DOI: 10.3945/ajcn.2009.28001. Epub 2009/09/16. PubMed PMID: 19759166; PubMed Central PMCID: PMCPMC2762163

[116] Arjmandi BH, Lucas EA, Khalil DA, Devareddy L, Smith BJ, McDonald J, et al. One year soy protein supplementation has positive effects on bone formation markers but not bone density in postmenopausal women. Nutrition Journal. 2005;**4**:8. DOI: 10.1186/1475-2891-4-8. Epub 2005/02/23. PubMed PMID: 15727682; PubMed Central PMCID: PMCPMC554088

[117] Levis S, Strickman-Stein N, Ganjei-Azar P, Xu P, Doerge DR, Krischer J. Soy isoflavones in the prevention of menopausal bone loss and menopausal symptoms: A randomized, double-blind trial. Archives of Internal Medicine. 2011;**171**(15):1363-1369. DOI: 10.1001/archinternmed.2011.330. PubMed PMID: 21824950

Chapter 3

Epidemiology and Management of Intracerebral Hemorrhage in Chile

Álvaro Soto, Marcelo Peldoza and Debora Pollak

Abstract

Intracerebral hemorrhage (ICH) is the second cause of stroke in Chile (23% of all strokes). The Araucanía Region has double the mortality rate by stroke compared with most regions in Chile. In developing countries like Chile, it is difficult to admit patients with ICH to the intensive care unit (ICU) for general neuroprotection and an aggressive lowering of blood pressure. The aim is to report the experience in the treatment of patients with ICH in a regional public hospital in Temuco, Chile. A convenience sample of the ICH consultations made during shift # 1 in the emergency room (ER) of the Hospital Dr. Hernán Henríquez Aravena between January 2016 and December 2018 was analyzed. There were 108 consultations for ICH in the period. The average age of the patients was 66.0 years (SD = 14.1). About 56.5% of the patients were male. Regarding the etiology, 70.4% were hypertensive, 18.5% were due to amyloid angiopathy, and 11.1% were for other causes. The implementation of 24/7 neurology shifts in the ER allowed us to reduce the evaluation time and improve the management of ICH patients. On the other hand, our hospital lacks a stroke unit, so ICH patients do not receive the standard care.

Keywords: intracerebral hemorrhage, stroke, epidemiology, management, Chile, hypertension, developing countries

1. Introduction

Stroke is the leading cause of disability and the second cause of death worldwide [1]. More than two thirds of the global burden of stroke occurs in developing countries, where the average age of patients is 15 years younger than in developed countries [2]. In the period 2000–2008, the total incidence rates in low- and middle-income countries exceeded the level of stroke incidence in high-income countries by 20% for the first time [3]. Latin America is experiencing an epidemiological transition toward older urban-dwelling adults that has led to a rise in cardiovascular risk factors and an increase in morbidity and mortality rates related to both stroke and myocardial infarction [4, 5]. According to the Global Burden of Disease 2013 Study (GBD 2013), stroke is the second cause of death in Latin America [6].

Stroke is also a serious public health problem in Chile. It is the leading cause of death in Chile, with a rate of 50.6 deaths per 100,000 inhabitants in 2011 [7]. Stroke accounted for 9% of all deaths in 2010 (8888 people) [8]. In addition, stroke is the first specific cause of disability-adjusted life years (DALY) in people older than 74 years and the fifth in those between 60 and 74 years [7]. 26,072 were hospitalized with the diagnosis of stroke in Chile in 2009 [8].

The prevalence of stroke in Chile, according to the National Health Survey (NHS) 2016–2017, is 2.6% in the general population and rises to 8.2% in ≥65 years [9]. A slight increase was observed when comparing the prevalence estimated in the 2009–2010 NHS, with 2.2 and 8.1%, respectively [10].

Intracerebral hemorrhage (ICH) is the second cause of stroke in Chile and represents approximately 23% of all strokes [11]. According to the Global Burden of Disease 2010 Study, the incidence of ICH in Chile is 46.9 per 100,000 person-years; the mortality is 22.36 per 100,000 person-years, and DALYs lost are 443.9 [12]. The comparison of the incidence, mortality, and DALYs between 1990 and 2010 is shown in **Table 1**.

The main source of information about the epidemiology of stroke comes from the Proyecto Investigación de Stroke en Chile: Iquique Stroke (PISCIS) Study conducted in Iquique in the north of Chile in 2000–2002 [11]. This study included 69 cases of first-ever ICH. Of these, 64 (92.7%) had spontaneous ICH. The mean age was 57.3 ± 17 years, and 62.3% of the subjects were male. The age-adjusted incidence rates were 13.8 (non-lobar) and 4.9 (lobar) per 100,000 person-years. Non-lobar ICH was more frequent in young men and lobar ICH in older women. The non-lobar-to-lobar ratio was similar to previous findings in Hispanics. Hypertension was more frequent in non-lobar ICH and in diabetes, while heavy drinking and antithrombotic use were more frequent in lobar ICH, but in none significantly. There was no association between location and prognosis [13]. In the PISCIS Study, the incidence rate per 100,000 was 27.6 for ICH. The case-fatality rate for incident ICH was 28.9 (17.7–44.8). The outcome at 6 months after the first-ever ICH was 33% of patients at mRankin 0–2, 28% at mRS 3–5, and 39% dead [11].

The INTERSTROKE (risk factors for ischemic and intracerebral hemorrhagic stroke in 22 countries) Study showed that hypertension, smoking, waist-to-hip ratio, diet, and alcohol intake were significant risk factors for ICH [14]. According to the National Health Survey (NHS) 2016–2017, 27.6% of the population in Chile has hypertension; 73.3% in the subgroup ≥65 years old; 12.3% with diabetes (30.6% in ≥65); 74.2% with overweight-obesity; 86.7% with physical inactivity; 11.7% with alcoholism; and 33.3% who smoke [9].

In relation to the in-hospital management of ICH in Chile, the percentage of patients admitted with ICH varies from 14% in a private neurological intermediate care unit to 34% of stroke cases in a public hospital in Santiago [15, 16].

The most important risk factor for ICH is age. Each advancing decade from 50 years of age is associated with a twofold increase in ICH incidence [17]. In other words, ICH is more common in the elderly (1.97 x for each 10-year increase) [18]. This is a very important issue because according to the 2017 Chilean National Census, 11.4% of the population is 65+ years old [19]. On the other hand, in the Araucanía Region, 12.6% of population is 65+ years old; this region is the second oldest after the Valparaíso Region (13.6% 65+ years old) [19].

Year	Incidence	Mortality	DALYs[a]
1990	58.26 (42.38–76·24)	43.21 (38.58–48.77)	884.19 (787.41–996.20)
2010	46.93 (35.24–61.38)	22.36 (19.41–26.57)	443.90 (385.72–519.42)

[a]*Disability-adjusted life-years.*

Table 1.
Age-standardized incidence and mortality per 100,000 person-years and DALYs lost per 100,000 people, for hemorrhagic stroke, in Chile in 1990–2010.

1.1 Intracerebral hemorrhage in the Araucanía region

The Araucanía Region has an area of 31,842.3 km^2, which represents 4.2% of the American and insular territory [20]. The region has a 17.2% poverty by income, twice the national rate (8.6%) [21]. It also has 9.9 years average schooling (11.1% national) and 29.1% rurality that it is the second at the national level [19]. The Temuco-Padre Las Casas (PLC) conurbation has approximately 360,000 inhabitants [19].

According to the 2017 Chilean National Census, 34.0% of those surveyed in the Araucanía Region stated they belonged to an indigenous or native group, a proportion significantly higher than the 12.8% registered nationally [19]. According to the National Socioeconomic Characterization Survey (CASEN) 2015, the indigenous population in Chile has worse socioeconomic indicators than nonindigenous [22]. For example, 18.3% of the indigenous population lives in poverty by income compared to 11% of the nonindigenous population; extreme poverty by income was 6.6 vs. 3.3%, respectively. Eighty-seven percent of the indigenous population is served in the public health system compared to 76.3% of the nonindigenous population [22]. There is evidence of a higher incidence of stroke among native populations and minorities [23, 24]. However, in a recent case-control study, we found no association between Mapuche ethnicity and stroke incidence. This study only included 16 patients with ICH [25].

The Araucanía Region, along with the Valparaíso, Maule, and Bío Bío regions, has double the mortality rate by stroke compared with the rest of the regions in Chile. Most of the increased risk is due to the prevalence of poverty, diabetes, sedentary lifestyle, and overweight [26]. Furthermore, according to the 2009–2010 NHS, the Araucanía Region has the highest prevalence of high systolic blood pressure compared to the other regions [10].

On the other hand, the incidence rate of stroke, calculated as a diagnosis of hospital discharge, in the period 2001–2010 in the Araucanía Sur Health Service, was 961.3 per 100,000 inhabitants/year [27].

1.2 Management of intracerebral hemorrhage in the HHHA

ICH is the fourth most frequent reason for neurological consultation in the emergency room (ER) of the Hospital Dr. Hernán Henríquez Aravena (HHHA) in Temuco, Chile, accounting for 4.5–7% of the care provided by the neurologist and 13.5–18.1% of stroke cases [28, 29].

The HHHA is located in the heart of the Temuco-PLC conurbation (360,000 inhabitants), about 670 kilometers south of Santiago de Chile. The HHHA has 730 beds, is the only hospital of high complexity in the Araucanía Region, and serves a beneficiary population of approximately 800,000 inhabitants [30]. The Araucanía Sur Health Service also has four medium-complexity hospitals (nodes) and eight low-complexity hospitals. The HHHA is also a referral center for neurological emergencies from the Araucanía Norte Health Service.

The HHHA neurology unit does not have its own service and depends on the internal medicine service. Our hospital lacks a stroke unit [30]. The hospital has two CT scanners and a MRI. There is an interventional neuroradiologist (MP) during daytime hours.

The HHHA has face-to-face neurologists 24/7 in the ER since July 2013 [29]. Patients with mild ICH (ICH score 0–1) are admitted to the internal medicine service [31]. Patients with severe ICH (ICH score 2–3) are admitted to the ICU. The ICU has 54 beds (18 with mechanical ventilation) for a population of about 800,000 inhabitants. Most patients with ICH stay a long time (24–48 h) in the ER waiting for a bed in the ICU. In these conditions it is very difficult to provide the standard care to these patients, including intensive blood pressure management and general neuroprotection. Based on the results of INTERACT2 and ATACH-2 studies, our target for systolic blood pressure in the first 48 h is less than 140 mmHg [32, 33]. Intravenous labetalol and nitroglycerin are the drugs more frequently used.

Another issue is the delay for the presentation of ICH patients. In a recent study, we estimated a median of 4 h and 45 mins (P_{25}–P_{75} = 3 h 13′–14 h 16′) for arrival to the ER. Just 17.4% of patients with ICH arrived in less than 3 h. In a chi-square test, the variables associated with a presentation in under 3 h were living in Temuco-PLC ($p < 0.01$), urban origin ($p = 0.02$), arrival by own car ($p = 0.032$), and severity (NIHSS ≥7) ($p < 0.01$). In a logistic regression model, only living in Temuco-PLC and severity were statistically significant with a combined odds ratio of 5.97 (95% CI = 3.23–11.04) [34].

The objective of this chapter is to report the experience in the treatment of patients with ICH in a regional public hospital in Temuco, Chile.

2. Material and methods

We performed a descriptive study of ICH in our hospital. A convenience sample of the consultations for ICH made during shift # 1 at the ER between January 2016 and December 2018 was analyzed. All patients were evaluated and diagnosed by AS in the ER. Due to the huge number of stroke patients diagnosed in the ER of the HHHA (about 220 cases of ICH per year), it was not possible to access to the clinical data of all patients with ICH in the period. We collected clinical, biodemographic, and imaging data. The radiological data were reviewed in all cases by a neuroradiologist (MP) unblinded to the clinical data. The cases were allocated to either spontaneous ICH or secondary to vascular malformation (arteriovenous malformation, cerebral saccular aneurysm, or cavernous angioma), tumor, or anticoagulants. On the other hand, the cases were classified according to one of two possible locations: supra- or infratentorial. Supratentorial hemorrhages included lobar, basal ganglia, and thalami locations. Infratentorial hemorrhages included the brainstem or cerebellum. The volume of the hematoma was calculated using the formula ABC/2, where A is the greatest diameter of the hematoma on the slice with the largest diameter, B is the diameter of the hematoma on the axis perpendicular to A, and C is the number of axial slices in which the hematoma is visible, multiplied by the slice thickness [35].

The continuous variables were described with measures of central tendency and dispersion, mean ± standard deviation (SD), and/or medians with percentiles 25–75 (P_{25}–P_{75}). The STATA 14.2 software was used for the data analysis.

3. Results

There were 108 consultations for ICH in the period. The average age of the patients was 66.0 years (SD = 14.1). 56.5% of the patients were male. The median NIHSS was 14 points. The median time to arrival to the ER was 4 h and 45 min. The median ICH score at admission was 1 point. Only 39.8% of patients were admitted in the ICU. The mortality at 30 days was 30.6%. This value was equivalent with in-hospital mortality. The clinical and biodemographic characteristics of the ICH patients are shown in **Table 2**. The radiological characteristics of patients are shown in **Table 3**.

Characteristics	Patients (N = 108)
Age (SD[a])	66.0 (14.1)
≥65 years (%)	55.6
≥80 years (%)	17.6
Male sex (%)	56.5
Mapuche ethnicity (%)	27.8
Rurality (%)	38.0
Temuco (%)	35.2
NIHSS[b] (median, P_{25}–P_{75})	14.0 (5–20)
Time to arrival (median, P_{25}–P_{75}) [min]	4 h 45′(3 h–14 h 21′)
Time to triage (median, P_{25}–P_{75})	8 min (5–15)
Time to evaluation (median, P_{25}–P_{75})	34 min (17–78)
ICH[c] score (median, P_{25}–P_{75})	1 (1–3)
ICH score (%)	0 = 23.1 1 = 32.7 2 = 15.4 3 = 14.4 4 = 12.5 5 = 1.9
Surgical hematoma evacuation (%)	3.1
EVD[d] placement (%)	2.1
Mortality (%)	
30-day	30.6
90-day	32.4
180-day	37.0
Destination (%)	Intensive care unit = 39.8 Internal medicine service = 32.4 Other hospital =24.1 Dye = 2.8 Discharge = 0.9

[a]*Standard deviation.*
[b]*National Institute of Health Stroke Scale.*
[c]*Intracerebral hemorrhage.*
[d]*External ventricular drain.*

Table 2.
Clinical and biodemographic characteristics of patients with intracerebral hemorrhage Hospital Dr. Hernán Henríquez Aravena, Temuco, Chile, in 2016–2018.

4. Discussion

In our study we found several similarities with many papers about ICH. In our series the mean age of the patients was 66 years old. The same was reported by Hemphill et al. (66 ± 15 years) but is higher than the age reported by Lavados et al. in the PISCIS Study (57.3 ± 17 years) [13, 31]. This difference could be explained because we used a convenience sample with cases that were not included consecutively. In our series we included about 1/6 of the ICH cases diagnosed in the ER of HHHA (about 220 cases/year). On the other hand, the locations found were similar to the findings of Hemphill et al.: 81.5%

Characteristics	Patients (N = 108)
Location 1	
Supratentorial	81.5
Infratentorial	18.5
Location 2	
Basal ganglia	32.4
Lobar	25.0
Thalamus	24.1
Cerebellum	10.2
Pons	8.3
Volume (cm^3, SD[a])	29.1 (37.6)
Intraventricular hemorrhage (%)	52.6
Etiology (%)	Hypertension = 70.4 Amyloid angiopathy = 18.5 Other = 11.1

[a]*Standard deviation.*

Table 3.
Radiological characteristics of patients with intracerebral hemorrhage Hospital Dr. Hernán Henríquez Aravena, Temuco, Chile, in 2016–2018.

supratentorial and 15.5% infratentorial [31]. We also found a 70.4% of the cases due to hypertension as the presumed cause. This is similar to the Hemphill study but higher than what was reported in the PISCIS Study [13, 31]. This difference in the results in comparison with our study can be explained because in the PISCIS Study, the patients were younger.

We found a 30-day mortality of 30.6% which is lower than the mortality reported by Hemphill in 2001 (45%) and similar to the 28.9% reported in Iquique, Chile [11, 31]. We also found a 6-month mortality of 37.0% which is similar to the 39% found in the PISCIS Study [11].

Unlike what was reported by Hemphill who found an association between mortality and age over 80 years, we found an association with age 65+ years ($p = 0.091$) [31]. In this sense we consider useful the modification in the ICH score proposed by Hegde et al. by reducing the age criteria by 10 years to prognosticate the disease better in populations belonging to developing countries like Chile [36].

Only 39.8% of our patients were admitted in the ICU. This reality is completely contrary to the clinical recommendations in developed countries. For instance, the American guideline for ICH management states that the initial monitoring and management of ICH patients should take place in an intensive care unit or dedicated stroke unit with physician and nursing neuroscience acute care expertise (Class I; Level of Evidence B) [37]. We can suppose that many patients die because they do not receive the care that the severity of their illness requires.

When presenting our results, we must emphasize that the HHHA does not have a specific infrastructure to attend to neurological patients, that is, a stroke unit. These units have demonstrated their cost-effectiveness in decreasing mortality and disability due to stroke [38]. In our situation, not all ICH patients are admitted to the ICU and complete 24–48 h of observation in the ER, being later hospitalized at the internal medicine service.

About 80% of the population in Chile is treated in the public health system [21]. Users of the public system have worse health indicators than users of the private health system [9]. On the other hand, it is expected that the incidence of ICH will increase significantly in our country due to the aging of the population and the poor control of cerebrovascular risk factors. This is why we see the need to have a stroke unit and/or a neurologic intermediate care unit in our hospital for the adequate management of patients with ICH. We also hope to set a Telestroke system with the future primary stroke centers in our region (Nueva Imperial, Pitrufquén, Lautaro, Villarrica, Victoria, and Angol hospitals). In short, we hope that the HHHA will become a comprehensive stroke center [39]. We also consider a priority to develop a better access in the detection and treatment of all the vascular risk factors mainly the control of hypertension. In the NHS 2003 in Chile, only 60% of the hypertensive knew their condition, 33% were being treated, and only 30% had normal values [40].

5. Conclusion

ICH is a common cause of consultation in our hospital, especially in older people. The implementation of 24/7 neurology shifts in the emergency room allowed us to reduce the evaluation time and to improve the management of ICH patients; however, it is still difficult to admit ICH patients to the ICU. We are aiming for a soon implementation of a stroke unit, so ICH patients receive a standardized care. It's a main priority to have better access to primary care prevention, diagnosis, and treatment in developing countries like ours.

Acknowledgements

This study is funded by the Universidad de La Frontera.

Conflict of interest

None.

Author details

Álvaro Soto[1,2,3*], Marcelo Peldoza[4,5] and Debora Pollak[1,2]

1 Facultad de Medicina, Departamento de Especialidades Médicas, Universidad de La Frontera, Temuco, Chile

2 Unidad de Neurología, Hospital, Dr. Hernán Henríquez Aravena, Temuco, Chile

3 Centro de Investigación en Epidemiología Cardiovascular y Nutricional (EPICYN), Universidad de La Frontera, Temuco, Chile

4 Departamento de Ciencias Preclínicas, Facultad de Medicina, Universidad de La Frontera, Temuco, Chile

5 Unidad de Neurorradiología Intervencional, Hospital, Dr. Hernán Henríquez Aravena, Temuco, Chile

*Address all correspondence to: alvaro.soto@ufrontera.cl

References

[1] Bonita R, Mendis S, Truelsen T, et al. The global stroke initiative. Lancet Neurology. 2004;**3**:391-393

[2] Truelsen T, Bonita R, Jamrozik K. Surveillance of stroke: A global perspective. International Journal of Epidemiology. 2001;**30**:S11-S16

[3] Feigin VL, Lawes CMM, Bennett DA, Barker-Collo SL, Parag V. Worldwide stroke incidence and early case fatality reported in 56 population-based studies: A systematic review. Lancet Neurology. 2009;**8**:355-369

[4] Avezum Á, Costa-Filho FF, Pieri A, Martins SO, Marin-Netoz JA. Stroke in latin America: Burden of disease and opportunities for prevention. Global Heart. 2015;**10**(4):323-331

[5] Ôunpuu S, Anand S, Yusuf S. The impending global epidemic of cardiovascular diseases. European Heart Journal. 2000;**21**:880-883

[6] Feigin VL, Mensah GA, Norrving B, et al. GBD 2013 stroke panel experts group. Atlas of the global burden of stroke (1990-2013): The GBD 2013 study. Neuroepidemiology. 2015;**45**:230-236

[7] Ministerio de Salud de Chile. Plan de Acción Ataque Cerebrovascular, 2a Edición [Internet]. 2014. Available from: http://www.redcronicas.cl/wrdprss_minsal/wp-content/uploads/2014/03/Plan-de-acci%C3%B3n-Ataque-Cerebrovascular-2014.pdf [Accessed: 23 June 2018]

[8] Ministerio de Salud de Chile. Guía clínica AUGE. Accidente Cerebrovascular Isquémico en personas de 15 años y más. Serie de las guías clínicas de MINSAL [Internet]. 2013. Available from: http://www.minsal.cl/portal/url/item/7222754637e58646e04001011f014e64.pdf [Accessed: 23 June 2018]

[9] Encuesta Nacional de Salud Chile 2016-2017 [Internet]. 2017. Available from: http://www.minsal.cl/wp-content/uploads/2017/11/ENS-2016-17_PRIMEROS-RESULTADOS.pdf [Accessed: 09 March 2018]

[10] Encuesta Nacional de Salud (ENS) Chile 2009-2010 [Internet]. 2010. Available from: http://www.redsalud.gov.cl/portal/url/item/99bbf09a908d3eb8e04001011f014b49.pdf [Accessed: 07 May 2013]

[11] Lavados PM, Sacks C, Prina L, Escobar A, Tossi C, et al. Incidence, 30-day case-fatality rate, and prognosis of stroke in Iquique, Chile: A 2-year community-based prospective study (PISCIS project). Lancet. 2005;**365**:2206-2215

[12] Krishnamurthi RV, Feigin VL, Forouzanfar MH, Mensah GA, Connor M, Bennett DA, et al. On behalf of the global burden of diseases, injuries, and risk factors study 2010 (GBD 2010) and the GBD stroke experts group*global and regional burden of first-ever ischaemic and haemorrhagic stroke during 1990-2010: Findings from the global burden of disease study 2010. The Lancet Global Health. 2013;**1**:e259-e281

[13] Lavados PM, Sacks C, Prina L, Escobar A, Tossi C, Araya F, et al. Incidence of lobar and non-lobar spontaneous intracerebral haemorrhage in a predominantly hispanic-mestizo population – The PISCIS stroke project: A community-based prospective study in Iquique, Chile. Neuroepidemiology. 2010;**34**:214-221

[14] O'Donell MJ, Xavier D, Liu L, et al. Risk factors for ischaemic and intracerebral haemorrhagic stroke in 22 countries (the INTERSTROKE study): A case-control study. Lancet. 2010;**376**:112-123

[15] Mellado P, Court J, Godoy J, Mery V, Barnett C, Andresen M, et al. Cerebrovascular disease in a neurologic intermediate care unit in Chile. Analysis of 459 consecutive patients. Revista Médica de Chile. 2005;**133**:1274-1284

[16] Nogales-Gaete J, Núñez L, Arriagada C, Sáez D, Figueroa T, Fernández R, et al. Clinical characterization of 450 patients with cerebrovascular disease admitted to a public hospital during 1997. Revista Médica de Chile. 2000;**128**(11):1227-1236

[17] Adeoye O, Broderick JP. Advances in the management of intracerebral Hemorrhage. Nature Reviews. Neurology. 2010;**6**:593-601

[18] Ferro JM. Update on intracerebral haemorrhage. Journal of Neurology. 2006;**253**:985-999

[19] Instituto Nacional de Estadísticas – Chile. Resultados Censo 2017 [Internet]. 2018. Available from: http://www.censo2017.cl/wp-content/uploads/2018/05/presentacion_de_la_segunda_entrega_de_resultados_censo2017.pdf [Accessed: 15 May 2018]

[20] Instituto Nacional de Estadísticas-Chile. Compendio Estadístico Región de La Araucanía [Internet]. 2017. Available from: http://www.inearaucania.cl/archivos/files/pdf/Ediciones%20Especiales/Compendio%202017.pdf [Accessed: 05 March 2019]

[21] Gobierno de Chile. Ministerio de Desarrollo Social, Encuesta de Caracterización Socioeconómica Nacional (Casen) 2017. Situación de pobreza [Internet]. 2017. Available from: http://observatorio.ministeriodesarrollosocial.gob.cl/casen-multidimensional/casen/docs/Resultados_pobreza_Casen_2017.pdf [Accessed: 05 March 2019]

[22] Gobierno de Chile. Ministerio de Desarrollo Social, Encuesta de Caracterización Socioeconómica Nacional (Casen) 2015 [Internet]. 2015. Available from: http://observatorio.ministeriodesarrollosocial.gob.cl/casen-multidimensional/casen/docs/CASEN_2015_Resultados_pueblos_indigenas.pdf. [Accessed: 13 November 2018]

[23] Stansbury JP, Jia H, Williams LS, et al. Ethnic disparities in stroke: Epidemiology, acute care, and Postacute outcomes. Stroke. 2005;**36**:374-387

[24] Trimble B, Morgenstern LB. Stroke in minorities. Neurologic Clinics. 2008;**26**:1177-1190

[25] Soto A, Morales G, Provoste R, Lanas F, Aliaga I, Pacheco D, et al. Association between Mapuche ethnicity and stroke: A case-control study. Journal of Stroke and Cerebrovascular Diseases. 2019;**28**(5):1311-1316

[26] Lavados PM, Díaz D, Jadue L, Olavarría VV, Cárcamo DA, Delgado I. Socioeconomic and cardiovascular variables explaining regional variations in stroke mortality in Chile: An ecological study. Neuroepidemiology. 2011;**37**:45-51

[27] Doussoulin A, Rivas R, Sabelle C. Egresos hospitalarios por enfermedad cerebrovascular en el período 2001-2010 en el Servicio de Salud Araucanía Sur. Revista Médica de Chile. 2016;**144**:571-576

[28] Soto A, Morales G, Vega C, Echeverría G, Colinas MB, Canales P, et al. Tiempos de atención de urgencias neurológicas en un hospital regional de alta complejidad. Revista Médica de Chile. 2018;**146**:885-889

[29] Soto A, Morales G, Pollak D, Jara V. Análisis de las consultas neurológicas en el Servicio de Urgencia de un hospital terciario. Revista Chilena de Neuro-Psiquiatría. 2016;**54**(2):93-101

[30] Soto A, Morales G, Grandjean M, Pollak D, Del Castillo C, García P, et al. Evolución del protocolo de trombolisis endovenosa en ataque cerebrovascular isquémico agudo: 4 años de experiencia en el Hospital Doctor Hernán Henríquez Aravena de Temuco- Chile. Revista Médica de Chile. 2017;**145**:468-475

[31] Hemphill JC, Bonovich DC, Besmertis L, Manley GT, Johnston SC. The ICH score: A simple, reliable grading scale for Intracerebral Hemorrhage. Stroke. 2001;**32**:891-897

[32] Anderson CS, Heeley E, Huang Y, Wang J, Stapf C, Delcourt C, et al. Rapid blood-pressure lowering in patients with acute intracerebral hemorrhage. The New England Journal of Medicine. 2013;**368**(25):2355-2365

[33] Qureshi AI, Palesch YY, Barsan WG, Hanley DF, Hsu CY, Martin RL, et al. ATACH-2 trial investigators and the neurological emergency treatment trials network. Intensive blood-pressure lowering in patients with acute cerebral Hemorrhage. The New England Journal of Medicine. 2016;**375**(11):1033-1043

[34] Soto A, Morales G, Echeverría G, Colinas MB, Canales P, Contreras D. Tiempos de llegada de pacientes con Ataque Cerebrovascular en un hospital regional de alta complejidad. Revista Médica de Chile. 2019. In press

[35] Kothari RU, Brott T, Broderick JP, et al. The ABCs of measuring Intracerebral Hemorrhage volumes. Stroke. 1996;**27**:1304-1305

[36] Hegde A, Menon G. Modifying the Intracerebral Hemorrhage score to suit the needs of the developing world. Annals of Indian Academy of Neurology. 2018;**21**(4):270-274

[37] Hemphill JC 3rd, Greenberg SM, Anderson CS, Becker K, Bendok BR, Cushman M, et al. Woo D; on behalf of the American Heart Association stroke council, council on cardiovascular and stroke nursing, and council on clinical cardiology. Guidelines for the management of spontaneous intracerebral hemorrhage: A guideline for healthcare professionals from the American Heart Association/ American Stroke Association. Stroke. 2015;**46**:2032-2060

[38] Stroke Unit Trialists'Collaboration. Organised inpatient (stroke unit) care for stroke. Cochrane Database of Systematic Reviews. 2007;(4. Art. No: CD000197). DOI: 10.1002/14651858. CD000197.pub2

[39] Soto A. Intravenous Thrombolysis for Acute Ischemic Stroke in a High Complex Regional Hospital. Rijeka: Intech. DOI: 10.5772/ intechopen.7954479

[40] Margozzini P, Rigotti A, Ferreccio C, et al. Hypertension and the cardiometabolic syndrome in Chile: A review of concepts and consequences for the developing world. Therapeutic Advances in Cardiovascular Disease. 2007;**1**:83-90

Chapter 4

Impact of Chronic Medical and Neuropsychiatric Illnesses on Quality of Life and Life Expectancy among Patients at the University of Port Harcourt Teaching Hospital (UPTH)

Aborlo Kennedy Nkporbu

Abstract

The prevalence of chronic medical and psychiatric diseases has continued to increase worldwide, and their consequences have remained a growing concern. Acting with a number of sociodemographic and clinical variables, they tend to affect quality of life (QOL) and onward to life expectancy. There is a direct relationship between QOL and life expectancy. Hypertension and diabetes mellitus acting with adverse environmental factors reset and overamplify the sympathetic outflow, and this may worsen hypertension and or cause depression, dysthymia, and anxiety disorders. Tuberculosis and HIV are two chronic infective medical conditions that equally negatively affect quality of life well-being and life expectancy. These chronic medical and psychiatric conditions have been associated with reduced QOL and life expectancy. The aim of this study, therefore, was to determine the impact of chronic medical and psychiatric disorders (HIV, tuberculosis, diabetes mellitus, hypertension, dysthymia, and GAD) on quality of life and life expectancy. Following ethical approval and informed consent from the participants, 40 subjects from each group of HIV, tuberculosis, diabetes mellitus, hypertension, anxiety, and dysthymia were studied using sociodemographic/clinical questionnaire and the WHOQOL-Bref. The data were analyzed using the SPSS version 20 statistical package. Confidence interval was set at 95% while P-value of less than 0.05 was considered statistically significant. There was reduced QOL on physical, psychological, and social relationship, environment domains, and general health facet, respectively. Chronic medical and psychiatric conditions may contribute to reduced QOL and life expectancy. Management of patients with these conditions should necessarily include attention to their QOL and well-being.

Keywords: chronic medical and psychiatric diseases, QOL, life expectancy, UPTH

1. Introduction

The prevalence of chronic medical and psychiatric diseases has continued to increase worldwide, and their consequences have remained a growing concern.

IntechOpen

Acting with a number of sociodemographic and clinical variables, they tend to negatively affect life expectancy through, among many other pathways, reducing quality of life. Despite healthcare improvements, there has been little evidence of benefit on life expectancy in people with chronic medical and mental disorders [1–3].

A number of sociodemographic and clinical variables may however serve as key determinants of quality of life and life expectancy. In Nigeria, these medical conditions have remained on the rise [4–9]. Studies have noted a prevalence of 10–15% for diabetes mellitus and 4.6% for HIV infection [4–6]. Worldwide, it was estimated that diabetes affected 285 million adults (20–79 years) in 2010, and this figure would likely increase to about 439 million adults by 2030 [6]. The World Health Organization (WHO) has also estimated that 2 billion people, almost a third of the world's population, have latent TB [7–9] which is one of the leading causes of mortality worldwide [9–11]. About 8 million people develop tuberculosis every year, and out of this number, some 3 million die of it, and over 95% are from developing countries [9–11].

The global prevalence of high blood pressures has been estimated to be between 10 and 15% of adult populations [12], which is also in line with the findings in Africa [13]. However, other studies have reported a worldwide prevalence of 15–30% in adults [14]. The prevalence of hypertension has increased from 11.2% in 1990 to 27.9% in 2010 in rural communities in the Niger Delta and 44.3% in urban Lagos [15–17]. Over 36 million people have contracted HIV infection worldwide, and over 16 million people are said to have died from the disease [4, 11]. The prevalence of cancer diseases, schizophrenia, and dementia have all continued to increase [18–21].

In terms of mode of acquisition, while hypertension, diabetes mellitus, cancer, schizophrenia, and dementia have a clear genetic component, in addition to adverse environmental factors [22, 23], HIV and tuberculosis are mainly acquired infections [24–26]. Furthermore, HIV infection, tuberculosis, and schizophrenia are associated with a high level of stigma and social discrimination [26–29], another strong determinant of the degree of psychological impact of these chronic conditions. It is also worthy of note that while severe emotional trauma can directly cause hypertension, diabetes, and schizophrenia [30–32], it can only predispose an individual to acquiring HIV due to poor sense of judgment, leading to sexual indiscretion and other risk-bearing practices [24–26]. In their late stages, HIV and tuberculosis infections can also cause dementia and mental disorders including schizophrenia-like illnesses [33–37].

The choice for their comparison was basically borne out of the observation that they all share some common features in terms of chronicity, with subsequent need for long-term medications, direct or indirect effects on the central nervous system (CNS) [25], high rate of mortality [18–21] and morbidity [32–38], and impact on emotion [33–38]. In addition, patients with these conditions need extensive education, attitudinal change, and coping and healthy lifestyle including diet and exercise [39–42]. The illnesses are equally similar in terms of complications in the central nervous system [21, 23]. Diabetic ketoacidosis, HIV and hypertensive encephalopathies, CNS disseminated tuberculosis, some metastasis to the brain cells, as well as the direct CNS impairment may all directly or indirectly affect the brain cell functions and cognitive ability. This in turn may cause altered sensorium, neuro-affections, neuro-deficits, cognitive impairment, and seizures in some cases. Furthermore, all conditions can directly alter neurotransmitter levels due to direct toxic effects on the brain cells (neurons) either from the viral cells or other opportunistic infections, disseminated tuberculosis to the CNS, hypertensive encephalopathy, or ketoacidotic complication, significantly disrupting relevant neurotransmissions. This may affect particularly the limbic apparatus, the center that regulates mood and controls emotions, anger, and rage.

It is equally important to note that baseline adverse psychosocial factors or psychological distress have been implicated as predictors of schizophrenia, hypertension and diabetes, or HIV infection [9–12, 43], through impairment of judgment in the later [24–26]. Also, certain environmental as well as socioeconomic factors have been identified to predispose to tuberculosis like living in an overcrowded environment [9–12, 43].

Studies have indicated that life expectancy decreases with each additional chronic condition [1–3]. A study remarked that a 67-year-old person will live on average 22.6 additional years in the absence of any chronic conditions while a 67-year-old person will live 7.7 fewer years and 17.6 fewer years with 5 chronic conditions and ≥10 chronic conditions, respectively [44]. The same study found that the average marginal decline in life expectancy was 1.8 years with each additional chronic condition ranging from 0.4 fewer years with the first condition to 2.6 fewer years with the sixth condition. These results are consistent by sex and race [44].

Another study using a sample of Medicare beneficiaries enrolled as of January 2008, of 21 different chronic conditions and about 1.4 million persons aged 67 and above, found that, on average, a 75-year-old American woman who has no chronic conditions will live 17.3 additional years to more than 92 years old [45]. Conversely, similar individual with five chronic conditions will live, on average, only to age 87, while an individual with 10 or more chronic conditions will survive only to age 80 [45]. Women tend to live longer than men, while white people live longer than black. Clearly, the nature and number of the chronic diseases are important for life expectancy. An individual with heart disease at age 67 is estimated to live an additional 21.2 years on average, while someone of the same age diagnosed with Alzheimer's disease is only expected to live 12 additional years [46]. Different medical conditions have different life expectancy, but this difference gradually decreases with age and more comorbid conditions [44–49].

Struggling with multiple chronic illnesses shortens life expectancy dramatically [45–49], and for older individuals, chronic or multiple chronic conditions equally threaten to reverse recent gains in average life spans. The medical advances and new technologies that have allowed sick people to live longer may not be able to keep up with the growing burden of chronic diseases. It is becoming very clear that preventing the development of additional chronic conditions and giving adequate treatment when they occur in the middle ages and the elderly could be the only way to continue to improve life expectancy. Violence and adverse childhood events are said to speed up aging, and life span continues to increase with each generation.

Concerns about premature mortality among people with chronic mental and medical disorders have been increasing [1–3]. Higher general and specific causes of mortality in all or specific age groups have been identified for people with serious medical and mental illnesses [18–21]. People with chronic and severe mental illness have lower life expectancies of between 13 and more than 30 years than the general population [50], and a loss of 8.8 life years (14.1 years for men and 5.7 years for women) was estimated by a study which compared people treated for SMI and the general population in Massachusetts, USA [2]. Also, using a nationwide hospital discharge registry, a study reported a wide difference in life expectancy at age 30 for the main mental disorder categories compared to the general population, particularly for functional psychosis other than schizophrenia/affective psychosis (15.9 years lost), substance abuse (15.6 years lost), and organic psychosis (14.8 years lost) for men and organic psychosis (22.6 years lost), mental retardation (14.7 years lost), and substance abuse (18.8 years lost) for women [46]. Emphasis has been made on management of suicide risk and physical illness, minimum polypharmacy, and improvement of accessibility to physical healthcare [51].

Social security and different forms of life insurance policies have greatly helped in stabilizing life expectancy; in spite of this, a growing number of beneficiaries with multiple chronic conditions still have reduced life expectancy. The burden and stress of chronic disease could erase decades of progress. Life expectancy in the USA is rising more slowly than in other parts of the developed world. Many blame the obesity epidemic and related health conditions for the worsening health of the American population.

Functional limitations, including difficulty walking across the room or preparing meals, and health problems, such as high blood pressure, cancer, and diabetes, also predicted greater odds of experiencing a fall for adults 65 and older. Previous research indicates that older African Americans were more likely to live in extended family households. The availability of assistance at home could help older adults avoid scenarios or behaviors that could lead to falls.

Medication side effect is another important factor influencing longevity [51–57]. People with chronic illness tend to stay compliant longer on medications with less side effect profiles. For example, many people living with HIV find it difficult to continue the treatment regimen due to the side effects [53, 58]. Hence, decreasing the toxicity and side effects of HIV drugs will increase longevity, as this will increase their life span by increasing the amount of time that patients can stay on the life-saving treatment regimen and also increase quality of life [52–54]. Similarly, the antituberculosis and anticancer drugs are known to have serious side effects that can impair drug compliance, thereby reducing the life span of the affected individual [55–57, 59–62]. Bringing new drugs to market is an essential part of increasing the life expectancy of young people with HIV, but lowering the drugs' toxicity may have even greater health benefits for all HIV patients [52, 54]. Some side effects, such as increased cardiovascular risk, also cause problems that directly contribute to premature mortality and reduced life.

In spite of the current drug toxicity levels, young people with HIV add nearly 2 years to their lives by initiating HIV treatment regimens soon after infection [52]. If a new drug has a low toxicity and is well-tolerated by the patients, then they are more likely to take it regularly so that it is as effective as possible, and this will add to their life span. Reduction in the toxicity of new drugs has been associated with increase in the patient's quality-adjusted life expectancy by as much as 11%, or more than 3 years [52]. "Quality-adjusted life years" and "quality-adjusted life expectancy" are measures that are used to determine the value of different medical actions. For example, a potentially life-saving drug that is highly toxic, causes so much discomfort, and leaves a patient debilitated would have a lower value than a life-saving drug that does not have such side effects. Furthermore, there exists the negative psychological impact of being on medications for a long time (and in some cases a lifetime), which is in turn detrimental on quality of life and longevity.

Infectious diseases are a significant health concern especially in developing countries, and this has significantly contributed to life expectancy [24–29]. Of particular importance is the interface between the immune system and invading bacteria or virus and the proteins that protrude through the outer cell membrane of the bacteria or virus. Because these outer membrane proteins are on the outside of the antigenic cells, they are visible to the human immune system and therefore are targeted by antibodies. Antibodies are so tuned to recognize the three-dimensional structure of outer membrane proteins that they can attach to them with lock-and-key specificity, thereby labeling the foreign bacteria cell for elimination.

Mutations in the genes that code for outer membrane proteins can produce changes in the protein's structure, and if the key then no longer fits the lock, the genetic mutations allow the bacterium or virus to evade recognition by antibodies [9, 25, 29]. The intense selective pressure on the disease-causing microorganism to survive the immune response, coupled with increased mutation rates, produces the incredibly fast rate of genomic change in infectious organisms. Mutations occur randomly throughout the genome, but because they help the microorganism avoid elimination by the immune system, changes in outer membrane proteins appear much more often than would be expected by chance in the surviving organism. Furthermore, this initial burst of mutations during the acute phase of infection allows the bacteria or virus to survive the host's immune response, and this helps to establish a chronic infection including HIV and tuberculosis. In addition, the already weakened or compromised immunity sets a vulnerable pace for repeated reinfections or other new infections [24–29]. These infections eventually become chronic and invariably affect longevity.

The chronicity of these medical illnesses, persistent and recurrent symptoms, impairment in functioning capacity, other adverse and enduring environmental psychosocial burdens, and even the thought of these can also in turn affect quality of life and subsequently lower life expectancy. Although HIV- and tuberculosis-infected and cancer-affected patients under antiretroviral (arbacire, nevirapine), antituberculosis drug (isoniazid and cycloserine), and anticancer medication (methotrexate) therapy infrequently suffer acute organic psychotic complications, the chronicity of the disease places them at greater risk for psychiatric comorbidity than the general population [63–77].

The terms "quality of life" refers to the physical, psychological, and social domains of health, viewed as distinct areas that are influenced by a person's experiences, beliefs, expectations, and perception [78], ("which is referred to here collectively as perceptions of health"). Two things are significant in the above definition: the first is the subjective nature of QOL, and second is the need for a Clinician to assess all those areas of life considered as having significant impact on QOL. Quality of life assessment measures changes in physical, functional, mental, and social health in order to evaluate the human and financial cost and benefits of new programs and interventions [78, 79].

Quality of life therefore is impaired in many ways by the individual's level of independence, social relationships, personal beliefs, and their relationship to salient features of the environment in addition to their physical health and psychological state. Quality of life also consists of fulfilling needs, meeting of social expectations, and assessing opportunities by using abilities. Abilities are impaired by ill health and worse still chronic medical illnesses [78, 79]. The services rendered by healthcare givers in mental health help to moderate social demands, supplement opportunities, and restore abilities. Quality of life can be altered by both the immediate and the long-term consequences of treatment especially the case of chronic illnesses [80]. Since 1948, when the WHO defined health as being not only the absence of disease and infirmity but also the presence of physical, mental, and social well-being, quality-of-life issues and well-being have taken the center stage in healthcare practices and research [78, 79, 81]. Several studies have shown that chronic medical and mental illnesses often impair or have negative impacts on the quality of life and subjective well-being of persons across a whole range of areas [80, 82–89].

There has been a growing interest during the past decades for assessing determinant factors of patients' health-related quality of life (HRQOL), especially in chronic diseases [78–89]. Diabetes mellitus, HIV, tuberculosis, hypertensive, cancer, schizophrenia, and dementia are some of these chronic diseases that involve people of all races and to some extent all ages. They are considered common chronic diseases in most countries, and their prevalence has continued to increase. Several studies have shown that chronic illnesses often impair or have negative impacts on the quality of life and subjective well-being of persons across a whole range of areas [80, 82–89].

Tuberculosis, HIV, and cancer diseases weaken patients' physical functioning and impair their quality of life and hence may affect life expectancy [82–85]. It has become important that TB and HIV control programs as well as cancer awareness and prevention programs at public health clinics design strategies to improve the quality of health and life of these patients. In patients with chronic diseases, all predicted domains of quality of life (QOL), including general health perceptions, somatic sensation, psychological health, spiritual well-being, and physical, social, and role functioning, all tend to be negatively affected [86–93]. Social stigmatization, isolation, pill burden, long duration of therapy, sexual dysfunction, loss of income, and fear were additional specific problems related to chronic medical conditions [80, 82–88]. Despite available curative therapy, TB and its treatment still have significant short- and long-term consequences on patients' QOL [82, 83, 86–93]. QOL has also been characterized as "the ultimate goal of all health interventions" [81].

2. Aim

The aim of this study, therefore, was to determine the impact of chronic medical and neuropsychiatric disorders (HIV, tuberculosis, diabetes mellitus, hypertension, schizophrenia, and dementia) on life expectancy and the role of quality of life and well-being.

3. Methodology

Following ethical approval and informed consent from the participants, 20 subjects from each group of HIV, tuberculosis, diabetes mellitus, hypertension, schizophrenia, and dementia were recruited based on the study's inclusion and exclusion criteria. The study group comprised patients already diagnosed by the consultant physicians at their respective specialty clinics at the University of Port Harcourt Teaching Hospital and on treatment and have been regular on follow-up at their respective outpatient clinics. Participants were recruited through a simple random sampling. Those recruited were within 30–40 years of age, whose illness duration was within 3–5 years. Thirty-five normal individuals (five for each medical condition) were selected also via simple random sampling from among staff of the hospital, matched for age and sex, as controls.

Both subjects and controls were administered the study's instruments including the sociodemographic/clinical questionnaire, WHO Composite International Diagnostic Interview (WHO CIDI), and the WHOQOL-Bref. The data were analyzed using the SPSS version 20 statistical package. The cohorts and control were followed up for clinic attendance, dropout, quality of life, death, and survival rates after 5 years. Confidence interval was set at 95%, while P-value of less than 0.05 was considered statistically significant (**Tables 1** and **2**).

4. Results

Variable	DM	Hypertension	HIV	TB	Cancer	Schizo	Dementia
Age							
30–39 years	9 (45%)	6 (30%)	12 (60%)	9 (45%)	7 (35%)	11 (55%)	0 (0%)
40–49 years	11 (55%)	14 (70%)	8 (40%)	11 (55%)	13 (65%)	9 (45%)	20 (100%)
Sex							
Female	12 (60%)	10 (50%)	10 (50%)	9 (45%)	10 (50%)	12 (60%)	11 (55%)
Male	8 (40%)	10 (50%)	10 (50%)	11 (55%)	10 (50%)	8 (40%)	9 (54%)
Marital status							
Married	10 (50%)	8 (40%)	8 (40%)	9 (45%)	12 (60%)	6 (30%)	14 (70%)
Divorce	2 (10%)	1 (5%)	2 (10%)	3 (15%)	2 (10%)	0 (0%)	0 (0%)
Separated	3 (15%)	3 (15%)	3 (15%)	3 (15%)	1 (5%)	5 (25%)	3 (15%)
Single	3 (15%)	4 (20%)	6 (30%)	4 (20%)	3 (15%)	9 (45%)	1 (5%)
Widowed	2 (10%)	4 (20%)	1 (5%)	1 (5%)	1 (5%)	0 (0%)	2 (10%)
Education							
Primary	2 (10%)	3 (15%)	3 (15%)	3 (15%)	2 (10%)	2 (10%)	2 (10%)
Secondary	8 (40%)	7 (35%)	4 (20%)	7 (35%)	6 (30%)	10 (50%)	5 (25%)
Tertiary	9 (45%)	8 (40%)	7 (35%)	7 (35%)	11 (55%)	7 (35%)	13 (65%)
None	1 (5%)	2 (10%)	1 (5%)	3 (15%)	0 (0%)	1 (5%)	0 (0%)
Occupation							
Managers	2 (10%)	1 (5%)	0 (0%)	0 (0%)	0 (0%)	0 (0%)	3 (15%)
Professionals	3 (15%)	1 (5%)	1 (5%)	0 (0%)	4 (20%)	0 (0%)	4 (20%)
Clerical support workers	5 (25%)	5 (25%)	4 (20%)	5 (25%)	4 (20%)	0 (0%)	2 (10%)
Service and sales workers	2 (10%)	4 (20%)	4 (20%)	5 (25%)	3 (15%)	3 (15%)	3 (15%)
Skilled agricultural forestry and fishery workers	4 (20%)	5 (25%)	2 (10%)	4 (20%)	2 (10%)	4 (20%)	2 (10%)
Craft and related trade workers	2 (10%)	1 (5%)	4 (20%)	3 (15%)	3 (15%)	4 (20%)	2 (10%)
Plant and machine operators and assemblers	1 (5%)	1 (5%)	3 (15%)	1 (5%)	2 (10%)	3 (15%)	1 (5%)
Elementary occupation	1 (5%)	1 (5%)	2 (10%)	2 (10%)	2 (10%)	6 (30%)	1 (5%)
Armed forces occupation	0 (0%)	1 (5%)	1 (5%)	0 (0%)	0 (0%)	0 (0%)	0 (0%)
Income							
Low	5 (25%)	4 (20%)	8 (40%)	9 (45%)	6 (30%)	11 (55%)	7 (35%)
Average	13 (65%)	14 (70%)	11 (55%)	10 (50%)	18 (90%)	9 (45%)	6 (30%)
High	2 (10%)	2 (10%)	1 (5%)	1 (5%)	1 (5%)	0 (0%)	7 (35%)

Table 1.
Sociodemographic variables of patients with diabetes, hypertension, HIV, tuberculosis, cancer schizophrenia, and dementia.

Medical/mental condition	Respondents	Domains of quality of life				
		Domain 1 (physical)	Domain 2 (psychological)	Domain 3 (social relationship)	Domain 4 (environment)	GHF
DM	Control	51.97 ± 14.77	56.20 ± 22.19	57.51 ± 26.13	52.01 ± 16.91	48.34 ± 22.44
	Subjects	44.98 ± 13.064	56.60 ± 24.914	48.06 ± 26.114	44.95 ± 14.831	47.98 ± 21.896
Hypertension	Subjects	45.98. ± 13.064	46.60 ± 24.914	53.06 ± 26.114	56.95 ± 14.831	50.98 ± 21.896
	Control	60.46 ± 12.788	61.05 ± 13.362	66.80 ± 21.378	59.62 ± 16.503	51.91 ± 23.319
HIV	Subjects	53.70 ± 10.103	48.67 ± 15.016	46.84 ± 21.032	50.33 ± 10.456	63.83 ± 20.349
	Control	66.36 ± 13.698	67.85 ± 27.870	57.09 ± 14.888	67.50 ± 24.102	66.61 ± 22.418
TB	Subjects	43.18 ± 12.044	44.60 ± 32.823	44.06 ± 37.141	45.85 ± 11.236	46.98 ± 23.493
	Control	61.26 ± 13.428	62.05 ± 11.122	65.71 ± 23.241	56.71 ± 15.412	52.78 ± 14.271
Schizo	Subject	41.22 ± 13.015	53.40 ± 22.813	46.16 ± 37.213	45.39 ± 14.622	43.15 ± 21.533
	Control	62.46 ± 12.644	59.06 ± 53.362	62.80 ± 27.378	57.62 ± 16.503	54.91 ± 23.100
HIV	Subject	49.23. ± 13.072	59.63 ± 24.152	52.06 ± 15.345	58.95 ± 14.511	53.15 ± 21.414
	Control	67.36 ± 17.614	68.85 ± 38.833	56.09 ± 17.755	63.50 ± 34.107	65.61 ± 29.411
Cancer	Subject	41.98 ± 83.062	42.60 ± 74.215	47.06 ± 76.877	45.95 ± 25.837	43.98 ± 84.819
	Control	66.46 ± 12.788	67.05 ± 13.362	66.74 ± 28.414	62.62 ± 19.411	64.91 ± 23.744
Dementia	Subject	41.70 ± 75.111	45.67 ± 31.075	45.84 ± 91.923	45.33 ± 17.411	47.83 ± 29.384
	Control	58.69 ± 13.17	57.28 ± 56.39	58.49 ± 53.44	54.33 ± 29.83	60.62 ± 34.71
Statistical analysis		P = 0.001	P = 0.001	P = 0.002	P = 0.004	P = 0.24

DM, diabetes mellitus; HPT, hypertension; HIV, human immunodeficiency virus; TB, tuberculosis; GHF, general health facet.

Table 2.
Quality of life using the WHOQOL-Bref of subjects and controls with diabetes mellitus, hypertension, HIV, tuberculosis, schizophrenia, cancer, and dementia.

5. Discussion

Young adults from 35 years and above predominated the study group. Similarly, younger age adults were more prevalent among the HIV, tuberculosis, and schizophrenia groups than the group with hypertension and cancer. The dementia group was composed of entirely younger age adults. Infectious diseases both sexually and nonsexually infections were predominantly represented by patients in early adulthood [4, 7]. Similarly, tuberculosis is an infective disease, and exposure to it may be more in adults who work in healthcare facilities and are mostly caregivers close to already infected persons [7–9]. Tuberculosis is also common in individuals whose immunity may have waned due to poor nutrition, alcohol use, stress, and other emotional illnesses, in most times occasioned by the medical illness, as well as other infective diseases [7–9]. HIV had the most of youngest population with age group 30–39 years forming 60%. This is in line with the earlier finding that HIV is more prevalent among young persons [24, 27]. Similarly, most cases of schizophrenia usually begin in early adulthood except paraphrenia that occurs among the elderly [21]. Hebephrenia, characteristically, one of the most early-onset and worst prognoses was excluded the same way as those with AIDS-defined illness that were excluded among the HIV group. Paranoid schizophrenia was the most common of the schizophrenic group.

Apart from the influence of genetic predisposition, diabetic mellitus and hypertension are largely a disease of lifestyle including unhealthy eating habits, alcohol intake, and sedentary lifestyle, which would be mostly displayed among adults [4, 13, 42, 48–50]. Hypertension is typically a disease of older adults but can occur in younger ages [12–14]. In this study, the majority of patients with hypertension were in the 40–49 age group. Cancers can occur in any age group [93–97], and those in this study ranged from lung, prostate, breast, cervical, blood, and bone cancers to cancers of the gastrointestinal tract. Dementia is generally a disease of advanced age but could present as presenile [98–102]. Even though all dementia participants sampled in this study were within the age group of 40–49, a majority were presenile. Patients with vascular dementia were excluded to reduce the influence of high blood pressure on this group. Overall, all the chronic illnesses in this study span from young to older ages. Therefore the choice of a fairly mid-age as inclusion criteria for respondents was to reduce the impact of extremes of age on the study variable particularly quality of life and life span. However, this may have contributed to the low mortality rate in this study.

In the study, female gender was slightly higher among all subjects with 56%. Females generally have better healthcare-seeking behavior and tend to have lower mortality rate at all ages [103]. So, even though hypertension has higher prevalence in community studies [6, 13, 17], most hospital-based studies show slightly higher prevalence in females. This is in addition to biological and cultural vulnerability of HIV infection [104]. Cancer was slightly higher among females, and this could also be due to the same reason that females report symptoms more readily than do males [103, 104].

The majority of the respondents were married. However, illnesses that have predominantly onset in early adulthood like schizophrenia or common among young adult like HIV had a high percentage of single people [21, 24, 27]. In addition to being predominant among young people, HIV and schizophrenia are associated with a high level of stigma [1, 77, 105], although stigma in the former seems to have reduced significantly over the years. Persons affected with schizophrenia particular early-onset type tend to have difficulty forming or sustaining relationship or even get married, and even those who are already married may face a high risk of separation due to fear of transmission.

A majority of the respondents were engaged in middle-class occupation followed by the lower cadre occupation. Hypertension and dementia which are conditions seen common in advance age were most common among the first-class occupations, while cancer, TB, HIV, and schizophrenia were more prevalent among those with lower cadre occupation. Similarly, about 95% of the respondents fall within low- and medium-income earners. A rewarding and satisfying job or occupation is key to good quality of life and by extension prolongs life span [78–84]. Because of the stigma associated with HIV, tuberculosis, schizophrenia, and to some extent cancer [1, 105, 106], there is reduced opportunity to secure sustained employment, so most of these individual settle for menial jobs and petty trading. Even those who had better jobs are sometimes laid off due to chronic illnesses especially in the private sector. Diabetes, hypertension and pre-senile dementia, though without risk from job discrimination, can be the cause of job dissatisfaction due to poor functioning and performance following disabling symptoms.

The cost of continuing treatment in chronic medical and mental disorders is usually huge on sufferers and their families [30]. This often is a major reason for poor drug compliance and in some cases treatment discontinuation [60–62] and may largely contribute to mortality [51, 58]. A majority of the respondents were low- and medium-income earners and may have difficulty in financing the management of their illness. This may reduce quality of life as well as life span. From the study, the dropout rates were high among dementia patients, followed by diabetes mellitus and schizophrenia. This may be due to financial difficulty or poor insight.

Quality of life among all respondents was below average on almost all domains and was statistically significant on all domains except the general health facet. The controls all had better quality of life as they all scored above average. Cancer and tuberculosis patients scored lowest on quality of life particularly both on physical and psychological domains and on social domains for the later. Apart from specific symptoms of these two medical conditions, they are usually associated with weight loss and extreme weakness, and as such the patient may have difficulty carrying out daily activities. In this case, physical domain of quality of life may be impaired. Psychological domains of quality of life of people suffering from chronic illnesses are usually first to be affected [64–73]. Cancers are associated with some level of stigma but not as high as seen in tuberculosis. The high level of stigma and discrimination associated tuberculosis usually affects the social domain of quality of life [82, 83, 106, 107]. Diabetes mellitus was the next with poor quality of life especially on physical and environmental domains, giving credence to a number of studies [107–110]. Diabetes presents with a lot of physical symptoms and risk of systemic damage and requires strict drug and dietary compliance. This may have contributed to the high rate of dropout and poor quality of life.

Hypertension had the best quality of life on all domains followed by HIV. These two conditions were equally found to have lower crude mortality rates. Generally, the study found that better quality of life directly correlated with higher person-years lived and inversely correlated with crude mortality rates (**Table 3**). Hypertension equally presents with a number of physical symptoms, the risk of systemic damage, and the need for strict drug compliance. However, it is not associated with any form of stigma, and a majority of respondents were average-income earners. The level of stigma and discrimination that was associated with HIV two decades ago has drastically reduced following massive public awareness. Also, governments of different countries and the WHO have continued to embark on different intervention strategies including free antiretroviral medications. These effects may have combined to reduce the financial and social burden including stigma and discrimination among HIV-positive individuals.

Condition	No. of alive at beginning of study	Average quality of life	No. of deaths	Dropout	Mean person-years lived	Crude mortality rate
Hypertension	20	49.84 ± 26.124	2 (10%)	1 (5%)	2.31	0.844
Diabetes	20	45.31 ± 71.012	3 (15%)	2 (10%)	2.15	1.395
Tuberculosis	20	41.92 ± 13.021	4 (20%)	0 (0%)	1.21	3.306
Schizophrenia	20	53.43 ± 22.081	1 (5%)	2 (10%)	3.42	0.292
HIV	20	47.51 ± 31.411	2 (10%)	1 (5%)	2.83	0.707
Cancer	20	42.61 ± 78.033	4 (20%)	1 (5%)	1.85	2.162
Dementia	20	48.19 ± 85.151	2 (10%)	3 (15%)	3.21	0.623
Control	35	67.22 ± 52.81	0 (0%)	1 (5%)	—	—

Table 3.
Record of death and survival of respondents.

Quality of life in people living with HIV (PLWHIV), dementia, schizophrenia, and tuberculosis was significantly affected, more on the psychological and social domains compared with diabetes mellitus, hypertension, and cancer diseases [92, 107–112]. This suggests that the stigma and social rejection associated with the communicable disease may play a significant role in the development of psychological illness. This also implies that even though psychological burden is equally common in the diabetes mellitus and hypertension and affecting quality of life generally, the presence of psychological burden and trauma that may be associated with PLWHIV and tuberculosis tended to have more severe negative impact on quality of life [65, 66]. Again, there is a possibility that there may have been existing psychological illnesses either undiagnosed or untreated that may have made them engage in risky sexual behaviors that may have made them vulnerable to infectious diseases.

The presence of symptoms of tuberculosis, cancers, hypertension, and diabetes mellitus alone appears to be more disabling than those in PLWHIV, dementia, and schizophrenia bearing in mind that acute cases were excluded. Moderate to severe cases of diabetes mellitus, hypertension, and tuberculosis cause more symptoms, and they are more disabling. This may account for the better quality of life among PLWHIV on the physical domain and the lower quality of life scores on both the physical and environment domains, among them, than the PLWHIV in this study. HIV not complicated with AIDS is most of the time symptom free or stable on medication, and this stability is often less sensitive to adverse environmental factors unlike in diabetes mellitus, hypertension, and tuberculosis where little adverse changes in the environment could affect profoundly the patients who had hitherto remained stable on medications [12, 22, 30, 39, 43, 52]. Such changes may include change in income level, employment, marital status (prolonged difficulty, disharmony, separation, divorce, or widowhood), and poor drug adherence with immediate exacerbation of symptoms.

A majority of the patients in all medical conditions fared well on most domains. The possible reasons are their focus on physical strength (e.g., evident physical health, absence of symptoms, ability to work around, available family support, and a strong religious belief) than on their weaknesses (e.g., social discrimination). On Domains 1 and 4, PLWHIV had better performance on quality of life, followed by hypertension and diabetes mellitus, while tuberculosis had the least in similar domains. Furthermore, PLWHIV also scored higher on GHF than the other medical conditions.

Chronic medical and mental diseases account for multiple burdens for patients, including the necessity to deal with pain, suffering, reduced quality of life, premature mortality, financial costs, and familial emotional trauma [12, 30, 32, 72, 73]. The risk factors for mental health problems among patients suffering from chronic medical illnesses are complex [72, 73, 113]. Usually, the more serious the somatic disease and symptoms are, the more probable it will be to be accompanied by mood and/or anxiety symptoms of variable severity [72, 73]; conditions arising after the somatic disease are diagnosed. In other words, even if those with dual diagnoses were excluded from the study, it clearly understood that most chronic medical conditions tend to be associated with some emotional disturbance. Failure to manage such mental health problems increases the patients' probability of suffering from complications, even lethal.

In chronic medical conditions, functionality may be severely impaired due to chronic psychogenic and somatic pain, frequent hospital admissions, and dependency from medical and nursing personnel. These are all markers of poor quality of life and well-being. It is important to mention that most of the mortalities in chronic medical conditions may not be due to the direct complication of the disease, rather a cumulative outcome of social and psychological dissatisfaction of the condition. Sufferers maintain the feeling that they have come to the end of the road and seek the easiest escape rooting out of the problem which is suicide [114–117]. In addition, research has pointed out a relationship between sustained emotional disturbance especially depression and reduced immunity. This may be worse among chronically ill patients, and this makes them more vulnerable to recurrent infections or reinfections. Good quality of life and well-being are a measure of satisfaction in major areas of life including mood stability and affording basic nutrition which will sustain immunity.

Most chronic illnesses particularly HIV, cancer, and tuberculosis in sub-Saharan Africa are classic examples of diseases with both medical and social dimensions, characterized by its close relation to poor socioeconomic conditions [27–31]. For instance, in tuberculosis, a higher risk of acquiring active disease occurs with alcoholism, smoking [48, 49], intravenous drug abuse [48, 49, 58, 74], diabetes mellitus, HIV infection, overcrowding, and other factors. The abovementioned risk factors are very prevalent among populations with reduced quality of life and well-being and increase risk of having HIV and progression from latent TB to active TB [77].

From the study, mortality was the highest among tuberculosis and cancer patients, followed by patients with diabetes mellitus. Schizophrenia had the lowest mortality after 5 years. There was no mortality among the control within the period. Correspondingly, mortally was the highest among the groups with the lowest quality of life. Quality of life is an indicator of total well-being and optimal health; therefore, if it is low, then it is an indication that the individual may not be enjoying good health. The finding among the diabetes group is in line with the earlier report that reduced life expectancy at age 15 by 1.3 years for men and 2.0 years for women in Canada [47] and a BMI of 40–45 kg/m^2 were associated with a 10-year reduction of life expectancy at age 35 compared to a BMI of 22.5–25 kg/m^2 [118]. Causes of mortality may be due to organ damage, complication of medication, systemic damage, or hemodynamic changes. Actual or direct causes of the deaths could not be ascertained as many of the deaths did not occur in the hospital. Mortality from tuberculosis and cancer tends to be high with a low rate of survival. Mortality in severe chronic mental illness is recognized to be raised, and underlying causes may be multiple. However, the death rate was lowest among schizophrenics in this study. This may be due to the fact the schizophrenics were predominantly young population and also because those with poor prognosis like hebephrenia and disorganized were excluded from the study and only paranoid and catatonic types which carry better prognosis were included. This finding however differs from the earlier finding that any psychiatric diagnosis was associated with a 65% higher than expected total mortality in a case register study in a British primary care cohort [119].

Coronary heart disease accounts for threefold elevated mortality in young adults with severe mental illness [120] and diabetes, while stroke is usually a complication of long-standing hypertension. Long-term antipsychotic use and adverse lifestyle choices (e.g., obesity, smoking, poor diet, illicit drug use, and physical inactivity) are implicated in increased risk of cardiovascular events in these populations [1, 51, 120–122] and clearly need higher levels of consideration in order to improve health and survival, as well as the better-known risks of suicide and violent deaths. The causal pathways between mental disorder and premature mortality are multiple, making life years lost an important outcome measure in this population.

Life expectancy is a commonly used indicator for how longevity may be impaired by specific long-term exposures (e.g., smoking, obesity, ethnicity, and socioeconomic status) or chronic conditions of ill health or risk (e.g., diabetes mellitus) [48, 49] and provides an alternative measure to determine the influences of different exposures for the purpose of highlighting premature mortality at younger ages in potentially vulnerable groups. It is therefore primarily a measure of *impact* and should be seen as complementary to more elaborate studies using measures of *effect*. As a measure, life expectancy analyses offer an important means of communicating impact on survival to policy makers. Current smoking is associated with around 4 to 5 life years lost for both genders [48, 49].

Clearly the mechanisms through which medical and mental disorders are associated with premature mortality will include the effects of these individual risk factors (e.g., smoking behavior, risk of diabetes, etc.) as well as other factors (such as risk of suicide or accidents and direct effects of mental distress on cardiovascular risk). Excess mortality associated with mental disorders has been demonstrated to be predominantly due to "natural" causes [18, 119, 120] although mental health service provision is often focused on preventing more rare outcomes of suicide and violent death [115–117]. If improving overall survival is to be considered as an alternative priority, much more efforts are clearly required to address the challenges of improving general health in people with mental disorders through medical services, socioeconomic support, and physical health promotion strategies [51].

It is important to note that a good number of psychosocial and clinical factors, like increased age, marital status (married), later age of onset of illness, education, employment, average to high monthly income, shorter duration of illness, longer duration of treatment, and emotional stability, may affect the quality of life and other outcome of the medical conditions. The implication of this is that these factors have to be addressed in the holistic management of these and indeed other chronic medical conditions.

6. Conclusion

The findings of this study support the impression that chronic medical conditions are associated with reduced quality of life, which, together with a number of sociodemographic and clinical factors, in turn affect life expectancy. The results support the call that the management of patients with these medical conditions should necessarily include attention to the mental health status of the sufferers.

7. Recommendations

Based on the findings of this study that chronic medical conditions are commonly associated with reduced quality of life and well-being, which further affect longevity, it becomes imperative that renewed efforts by government, aimed at both primary and secondary prevention, be intensified for these chronic medical conditions.

Author details

Aborlo Kennedy Nkporbu
Department of Neuropsychiatry, University of Port Harcourt Teaching Hospital, Port Harcourt, Nigeria

*Address all correspondence to: nakpigi2008@yahoo.com

References

[1] Gray R, Hardy S, Anderson KH. Physical health and severe mental illness: If we don't do something about it, who will? International Journal of Mental Health Nursing. 2009;**18**: 299-300

[2] Phillips RS, Wenger NS, Teno J, et al. Choices of seriously ill patients about cardiopulmonary resuscitation: Correlates and outcomes. The American Journal of Medicine. 1996;**100**:128-137

[3] Hagerty RG, Butow PN, Ellis PM, et al. Communicating prognosis in cancer care: A systematic review of the literature. Annals of Oncology. 2005;**16**: 1005-1053

[4] Centres for Disease Control and Prevention. HIV/AIDS Surveillance Report. 2004;**12**(1):1-42

[5] Shaw JE, Sicree RA, Zimmet PZ. Global estimates of the prevalence of diabetes for 2010 and 2030. Diabetes Research and Clinical Practice. 2010; **87**(1):4-14

[6] Unachukwu CN, Agomoh DI, Alasia DD. Pattern of non-communicable diseases among medical admissions in Port Harcourt, Nigeria. Nigerian Journal of Clinical Practice. 2008;**11**(1):14-17

[7] Kochi A. The global tuberculosis situation and the new control strategy of the World Health Organization. Tubercle. 1991;**72**(1):1-6

[8] World Health Organization. Global Tuberculosis Control. Geneva, Switzerland: World Health Organization; 2010. Available from: http://www.who.int/tb/publications/global_report/en/index.html

[9] Dye C, Scheele S, Dolin P, Pathania V, Raviglione MC. Global burden of tuberculosis: Estimated incidence, prevalence, and mortality by country. Journal of the American Medical Association. 1999;**282**(7):677-686

[10] Gwatkin D, Guillot M, Heuveline P. The burden of disease among the global poor. Lancet. 2000;**354**:586-589

[11] Murray CJ, Lopez AD. Mortality by cause for eight regions of the world. Global burden of disease. Lancet. 1997; **349**:1269-1276

[12] Kearney PM, Whelton M, Renoids K, Muntner P, Whenton KP. Global burden of hypertension: Analysis of worldwide data. Lancet. 2005;**365**:217-223

[13] Erhun WO, Olayiwola G, Agbani EO, Omotoso NS. Prevalence of hypertension in a University Community in South Western Nigeria. African Journal of Biomedical Research. 2005;**8**:15-19

[14] Van de Sande MAB, Bailey R, Faal H, Banya WA, Dolin P, Nyan OA, et al. Nationwide prevalence study of hypertension and related non-communicable diseases. Tropical Medicine & International Health. 1997; **2**:1039-1048

[15] Briganti EM, Shaw JE, Chadban SJ, Zimmet PZ, Welborn TA, McNeil JJ, et al. Untreated hypertension among Australian adults: The 1999-2000 Australian diabetes, obesity and lifestyle study. The Medical Journal of Australia. 2003;**179**:135-139

[16] Adefuye BO, Adefuye PO, Oladepo TO, Familoni OB, Olurunga TO. Prevalence of hypertension and other cardiovascular risk factors in an African urban, sub-urban religious community. Nigerian Medical Practitioner. 2009;**55**(1-2):4-8

[17] Wokoma FS, Alasia DD. Blood pressure pattern in Barako: A rural

community in Rivers State, Nigeria. Nigerian Health Journal. 2011;**11**:813

[18] Brown S, Inskip H, Barraclough B. Causes of the excess mortality of schizophrenia. The British Journal of Psychiatry. 2000;**177**:212-217

[19] Osborn DP, Levy G, Nazareth I, Petersen I, Islam A, et al. Relative risk of cardiovascular and cancer mortality in people with severe mental illness from the United Kingdom's General Practice Research Database. Archives of General Psychiatry. 2007;**64**:242-249

[20] Osborn D, Levy G, Nazareth I, King M. Suicide and severe mental illnesses. Cohort study within the UK general practice research database. Schizophrenia Research. 2008;**99**: 134-138

[21] McGrath J, Saha S, Chant D, Welham J. Schizophrenia: A concise overview of incidence, prevalence, and mortality. Epidemiologic Reviews. 2008;**30**:67-76

[22] Pickering TG, Devereux RB, James GD, et al. Environmental influences on blood pressure and the role of job strain. Journal of Hypertension. Supplement. 1999;**341**:120-124

[23] Oparil S, Zaman MA, Calhoun DA. Pathogenesis of hypertension. Annals of Internal Medicine. 2003;**139**(9):761-776

[24] Windle M. The trading of sex for money or drugs, sexually transmitted diseases (STDs) and HIV related risk behaviors among polysubstance use and alcoholic inpatients. Drug and Alcohol Dependence. 1997;**49**:33-38

[25] Johnson RT. In: Johnson A, editor. HIV in Viral Infections of the Nervous System. 2nd ed. PA, USA: Lippincott-Raven; 1998. pp. 287-331

[26] Center for Disease Control and Prevention. Prevention and treatment of tuberculosis patients infected with human immunodeficiency virus: Principles of therapy and revised recommendations. Morbidity and Mortality Weekly Report. 1998;**47** (RR-20):1-51

[27] Bharat S, Aggleton P. Facing the challenge: Household responses to AIDS in Mumbai, India. AIDS Care. 1999;**11**: 31-44

[28] Bird ST, Bogart LM, Delahanty DL. Health-related correlates of perceived discrimination in HIV care. AIDS Patient Care STDS. January 2004;**18**(1):19-26

[29] Courtwright A, Turner AN. Tuberculosis and stigmatization: Pathways and interventions. Public Health Reports. 2010;**125**(Supplement 4): 34-42

[30] Esteghamati A, Khalilzadeh O, Anvari M, Meysamie A, Abbasi M, Forouzanfar M, et al. The economic costs of diabetes: A population-based study in Tehran Iran. Diabetologia. 2009;**52**(8):1520-1527

[31] Rogacheva MG. Social aspects in tuberculosis among mental patients. ProblemyTuberkuleza. 2002;**10**:13-16

[32] World Health Organization. Burden: Mortality, morbidity and risk factors. Global Status Report on Non-communicable Diseases. 2010. Available from: http://www.who.int/nmh/publications/ncd_report_chapter_1.pdf [Accessed: 01 October 2012]

[33] Issa BA, Yussuf AD, Kuranga SI. Depression comorbidity among patients with tuberculosis in a university teaching hospital outpatient clinic in Nigeria. Mental Health in Family Medicine. 2009;**6**:133-138

[34] Deribew A, Tesfaye M, Hailmichael Y, et al. Common mental disorders in TB/HIV co-infected patients in Ethiopia. BMC Infectious

Diseases. 2010;**10**:201. DOI: 10.1186/1471-2334-10-201

[35] Chaudhry R, Mishra P, Mishra J, Parminder S, Mishra BP. Psychiatric morbidity among diabetic patients: A hospital-based study. Industrial Psychiatry Journal. 2011;**19**(1):47-49

[36] De Ornelas Maia AC, Braga Ade A, Brouwers A, Nardi AE, De Oliveira e Silva AC. Prevalence of psychiatric disorders in patients with diabetes types 1 and 2. Comprehensive Psychiatry. 2012;**53**(8):1169-1173

[37] Johnson J, Williams J, Rabkin J, et al. Axis 1 psychiatric symptomatology associated with HIV infection and personality disorder. The American Journal of Psychiatry. 1995;**152**:551-554

[38] Myers H, Durvasula B. Psychiatric disorders in African, American men and women living with HIV/AIDS. Cultural Diversity and Ethnic Minority Psychology. 1999;**5**:249-262

[39] Grange JM, Festenstein F. The human dimension of tuberculosis control. Tubercle and Lung Disease. 1993;**74**(4):219-222

[40] Lerman I, Lozano L, Villa AR, et al. Psychosocial factors associated with poor diabetes self-care management in a specialized center in Mexico City. Biomedicine & Pharmacotherapy. 2004;**58**:566-570

[41] Lin EH, Katon W, Von Korff M, et al. Relationship of depression and diabetes self-care, medication adherence, and preventive care. Diabetes Care. 2004;**27**:2154-2160

[42] Peltzer K, Louw J, Mchunu G, Naidoo P, Matseke G, Tutshana B. Hazardous and harmful alcohol use and associated factors in tuberculosis public primary care patients in South Africa. International Journal of Environmental Research and Public Health. 2012;**9**(9):3245-3257

[43] Rajeswari R, Balasubramanian R, Muniyandi M, Geetharamani S, Thresa X, Venkatesan P. Socio-economic impact of tuberculosis on patients and family in India. The International Journal of Tuberculosis and Lung Disease. 1999;**3**(10):869-877

[44] Danaei G, Rimm EB, Oza S, Kulkarni SC, Murray CJ, et al. The promise of prevention: The effects of four preventable risk factors on national life expectancy and life expectancy disparities by race and county in the United States. PLoS Medicine. 2010;**7**:e1000248

[45] Clarke CA, Miller T, Chang ET, Yin D, Cockburn M, et al. Racial and social class gradients in life expectancy in contemporary California. Social Science & Medicine. 2010;**70**:1373-1380

[46] Hannerz H, Borga P, Borritz M. Life expectancies for individuals with psychiatric diagnoses. Public Health. 2001;**115**:328-337

[47] Sikdar KC, Wang PP, MacDonald D, Gadag VG. Diabetes and its impact on health-related quality of life: A life table analysis. Quality of Life Research. 2010;**19**:781-787

[48] Ozasa K, Katanoda K, Tamakoshi A, Sato H, Tajima K, et al. Reduced life expectancy due to smoking in large-scale cohort studies in Japan. Journal of Epidemiology. 2008;**18**:111-118

[49] Tamakoshi A, Kawado M, Ozasa K, Tamakoshi K, Lin Y, et al. Impact of smoking and other lifestyle factors on life expectancy among Japanese: Findings from the Japan collaborative cohort (JACC) study. Journal of Epidemiology. 2010;**20**:370-376

[50] Colton CW, Manderscheid RW. Congruencies in increased mortality rates, years of potential life lost, and causes of death among public mental

health clients in eight states. Preventing Chronic Disease. 2006;**3**:A42

[51] Auquier P, Lancon C, Rouillon F, Lader M, Holmes C. Mortality in schizophrenia. Pharmacoepidemiology and Drug Safety. 2006;**15**:873-879

[52] Rabkin JG, Ferrando S, Lin SH, Sewell M, McElhiney M. Psychological effects of HAART: A 2-year study. Psychosomatic Medicine. 2000;**62**: 413-422

[53] Ellen SR, Judd FK, Mijch AM, Cockram A. Secondary mania in patients with HIV infection. Australian and New Zealand Journal of Psychiatry. 1999;**33**:353-360

[54] Barllett J. Addressing the challenges of adherence. Journal of Acquired Immune Deficiency Syndromes. 2002; **29**(Suppl 1):S-SO

[55] Prasad R, Garg R, Verma SK. Isoniazid- and ethambutol-induced psychosis. Annals of Thoracic Medicine. 2008;**3**(4):149-151

[56] Martin SJ, Bowden FJ. Ethambutol toxicity manifesting as acute onset psychosis. International Journal of STD and AIDS. 2007;**18**(4):287-288

[57] Kennedy NA, Oluwaseun A, Denis AA, Chukwuemeka SP. Cycloserine induced-psychosis in a 22-year old male pharmacy student: A case report. American Journal of Psychiatry and Neuroscience. 2016;**1**:124-126

[58] Chang CK, Hayes RD, Perera G, Broadbent M, Fernandes AC, et al. All-cause mortality among people with serious mental illness (SMI), substance use disorders, and depressive disorders in southeast London: A cohort study. BMC Psychiatry. 2010;**10**:77

[59] Prasad CE, Krishnamurthy K, Murthy KJR. Psychiatric disorders in patients receiving anti-tuberculosis drugs. Indian Journal of Psychiatry. 1985;**27**(4):311-314

[60] Bagchi S, Ambe G, Sathiakumar N. Determinants of poor adherence to anti-tuberculosis treatment in Mumbai, India. International Journal of Preventive Medicine. 2010;**1**(4):223-232

[61] Manoharam E, John KR, Joseph A, Jacob KS. Psychiatric morbidity, patients' perspectives of illness and factors associated with poor medication compliance among the tuberculous in Vellore, south India. The Indian Journal of Tuberculosis. 2001;**48**:77-80

[62] Erhabor GE, Aghanwa HS, Yusuph M, Adebayo RA, Arogundade FA, Omidiora A. Factors influencing compliance in patients with tuberculosis on directly observed therapy at Ile-Ife, Nigeria. East African Medical Journal. 2000;**77**(5):235-239

[63] Sulehri MA, Dogar IA, Sohail H, et al. Prevalence of depression among tuberculosis patients. Annals of Punjab Medical College. 2010;**4**:133-137

[64] Jonas BS, Franks P, Ingram DD. Are symptoms of anxiety and depression risk factors for hypertension? Longitudinal evidence from the National Health and Nutrition Examination Survey I Epidemiologic Follow-up Study. Archives of Family Medicine. 1997;**6**:43-49

[65] Katze S, Nevid JS. Risk factors associated with posttraumatic stress disorder symptomatology in HIV-infected women. AIDS Patient Care and STDs. 2005;**19**(2):110-120

[66] Olley BO, Zeier MD, Seedat S, Stein DJ. Posttraumatic stress disorder among recently diagnosed patients with HIV/AIDS in South Africa. AIDS Care. 2005;**17**(5):550-557

[67] Peltzer K, Naidoo P, Matseke G, Louw J, McHunu G, Tutshana B.

Prevalence of post-traumatic stress symptoms and associated factors in tuberculosis (TB), TB retreatment and/ or TB-HIV co-infected primary public health-care patients in three districts in South Africa. Psychology Health and Medicine. 2013;**18**(4):387-397

[68] Engum A. The role of depression and anxiety in onset of diabetes in a large population-based study. Journal of Psychosomatic Research. 2007;**62**(1): 31-38

[69] Jaggarajamma K, Ramachandran R, Charles N, Chandrasekaran V, Muniyandi M, Ganapathy S. Psycho-social dysfunction: perceived and enacted stigma among tuberculosis patients registered under revised national tuberculosis control programme. The Indian Journal of Tuberculosis. 2008;**55**(4):179-187

[70] Guruprasad KG, Niranjan MR, Ashwin S. A study of association of depressive symptoms among the type 2 diabetic outpatients presenting to a tertiary care hospital. Indian Journal of Psychological Medicine. 2012;**34**:30-33

[71] Poongothai S, Anjana RM, Pradeepa R, Ganesan A, Umapathy N, Mohan V. Prevalence of depression in relation to glucose intolerance in urban south Indians: The Chennai Urban Rural Epidemiology Study (CURES-76). Diabetes Technology & Therapeutics. 2010;**12**:989-994

[72] Katon WJ. Epidemiology and treatment of depression in patients with chronic medical illness. Dialogues in Clinical Neuroscience. 2011;**13**:7-23

[73] Katon W, Lin EH, Kroenke K. The association of depression and anxiety with medical symptom burden in patients with chronic medical illness. General Hospital Psychiatry. 2007;**29**:147-155

[74] Crosby G, Stall R, Paul J, Barrett D. Substances use and HIV risk profile of gay/bisexual males who drop substance abuse treatment. AIDS Education and Prevention. 2000;**12**:38-48

[75] Black SA. Increased health burden associated with comorbid depression in older diabetic Mexican Americans. Results from the Hispanic Established Population for the Epidemiologic Study of the Elderly survey. Diabetes Care. 1999;**22**:56-64

[76] Kaholokula JK, Haynes SN, Grandinetti A, Chang HK. Biological, psychosocial, and sociodemographic variables associated with depressive symptoms in persons with type 2 diabetes. Journal of Behavioral Medicine. 2003;**26**:435-458

[77] Ogunrin AO, Odiase FE, Ogunniyi A. Reaction time in patients with HIV/AIDS and correlation with CD4 count: A case-control study. Transactions of the Royal Society of Tropical Medicine and Hygiene. 2007; **101**(5):517-522

[78] Levine S, Croog S. What constitutes quality of life? A conceptualization of the dimensions of life quality in healthy population and patients with cardiovascular disease. In: Wenger N, Mattson ME, Furgerg CD, Elinson J, editors. Assessment of Quality of Life in Clinical Trials of Cardiovascular Therapies. New York: Le Jacq; 1998. pp. 46-58

[79] Robinson S, Young T, Roos L. Estimating the burden of disease. Comparing administrative data and self-reports. Medical Care. 1997;**35**(9): 932-947

[80] Testa MA, Hollenberg Anderson RA, Williams GH. Assessment of quality of life by patient and spouse during antihypertensive therapy with atenolol and nifedipine gastrointestinal therapeutic system. American Journal of Hypertension. 1991;**4**:363-373

[81] Concepts of health-related quality of life. In: Patrick DL, Erickson P, editor. Health Status and Health Policy: Quality of Life in Health Care Evaluation and Resource Allocation. New York: Oxford University Press; 1993. pp. 76-112

[82] Louw J, Peltzer K, Naidoo P, Matseke G, McHunu G, Tutshana B. Quality of life among tuberculosis (TB), TB retreatment and/or TB-HIV co-infected primary public health care patients in three districts in South Africa. Health and Quality of Life Outcomes. 2012;**10**:77. DOI: 10.1186/1477-7525-10-77

[83] Weis SE, Pasipanodya JG. Measuring health-related quality of life in tuberculosis: A systemic review—Response. Health and Quality of Life Outcomes. 2010;**8**:7. DOI: 10.1186/1477-7525-8-7

[84] Jia H, Uphold CR, Wu S, Reid K, Findley K, et al. Health-related quality of life among men with HIV infection: Effects of social support, coping and depression. AIDS Patient Care and STDs. 2004;**18**:594-603

[85] Ghanbari A, Yekta ZP, Roushan ZA, Lakeh NM. Assessment of factors affecting quality of life in diabetic patients in Iran. Public Health Nursing. 2005;**22**(4):311-322

[86] Haririan H, Moghadasian S, Aghajanlou A. Quality of life and its dimensions in diabetic patients referred to diabetes center of Tabriz medical university. Iranian Journal of Diabetes and Lipid Disorders. 2009;**9**(2):152-160

[87] Brown J, Capocci S, Smith C, Morris S, Abubakar I, Lipman M. Health status and quality of life in tuberculosis. International Journal of Infectious Diseases. 2015;**32**:68-75

[88] Aggarwal AN, Gupta D, Janmeja AK, Jindal SK. Assessment of health-related quality of life in patients with pulmonary tuberculosis under programme conditions. The International Journal of Tuberculosis and Lung Disease. 2013;**17**(7):947-953. DOI: 10.5588/ijtld.12.0299

[89] Rajeswari R, Muniyandi M, Balasubramanian R, Narayanan PR. Perceptions of tuberculosis patients about their physical, mental and social well-being: A field report from south India. Social Science & Medicine. 2005;**60**(8):1845-1853. DOI: 10.1016/j.socscimed.2004.08.024

[90] Adeyeye OO, Ogunleye OO, Coker A, Kuyinu Y, Bamisile RT, Ekrikpo U, et al. Factors influencing quality of life and predictors of low quality of life scores in patients on treatment for pulmonary tuberculosis: A cross sectional study. Journal of Public Health in Africa. 2014;**5**(2):88-92. DOI: 10.4081/jphia.366. Available from: https://www.publichealthinafrica.org/index.php/jphia/article/view/366/156

[91] Balgude A, Sontakke S. Study of impact of antitubercular therapy on quality of life. Indian Journal of Medical Sciences. 2012;**66**(3-4):71-77

[92] Wexler DJ, Grant RW, Wittenberg E, et al. Correlates of health-related quality of life in type 2 diabetes. Diabetologia. 2006;**49**:1489-1497

[93] Brustugun OT, Moller B, Helland A. Years of life lost as a measure of cancer burden on a national level. British Journal of Cancer. 2014;**111**:1014-1020. DOI: 10.1038

[94] Rachet B, Woods LM, Mitry E, Riga M, Cooper N, Quinn MJ, et al. Cancer survival in England and Wales at the end of the 20th century. British Journal of Cancer. 2008;**99**(Suppl 1):S2-S10

[95] Rutherford MJ, Andersson TML, Møller H, Lambert PC. Understanding the impact of socioeconomic differences

in breast cancer survival in England and Wales: Avoidable deaths and potential gain in expectation of life. Cancer Epidemiology. 2015;**39**:118-125

[96] Rutherford MJ, Hinchliffe SR, Abel GA, Lyratzopoulos G, Lambert PC, Greenberg DC. How much of the deprivation gap in cancer survival can be explained by variation in stage at diagnosis: An example from breast cancer in the East of England. International Journal of Cancer. 2013; **133**(9):2192-2200

[97] Siemerink EJM, Hospers GAP, Mulder NH, Siesling S, van der Aa MA. Disparities in survival of stomach cancer among different socioeconomic groups in north-east Netherlands. Cancer Epidemiology. 2011;**35**:413-416

[98] Singh GK, Williams SD, Siahpush M, Mulhollen A. Socioeconomic, rural-urban, and racial inequalities in us cancer mortality: Part I —All cancers and lung cancer and part II—Colorectal, prostate, breast, and cervical cancers. Journal of Cancer Epidemiology. 2011;**2011**:107497

[99] Dewey EM, Saz P. Dementia cognitive impairment and mortality in persons aged 65 and over living in a community a systematic review of literature. International Journal of Geriatric Psychiatry. 2001;**16**:751-761

[100] Agüero-Torres H, Fratiglioni L, Guo Z, Viitanen M, Winblad B. Mortality from dementia in advanced age: A 5-year follow-up study of incident dementia cases. Journal of Clinical Epidemiology. 1999;**52**:737-743

[101] Kammoun S, Gold G, Bouras C, Giannakopoulos P, McGee W, Herrmann F, et al. Immediate causes of death of demented and non-demented elderly. Acta Neurologica Scandinavica. Supplementum. 2000;**176**:96-99

[102] Mitchell SL, Kiely DK, Hamel MB, Park PS, Morris JN, Fries BE. Estimating prognosis for nursing home residents with advanced dementia. JAMA. 2004; **291**:2734-2740

[103] Nor K, Tlou S, Norr J. The threat of AIDS for women in developing countries. In: Cohen F, Durham JD, editors. Women Children and HIV. New York: Springer Publishing Co; 1993. p. 263

[104] Gibson N, Cave A, Doering D, Ortiz L, Harms P. Socio-cultural factors influencing prevention and treatment of tuberculosis in immigrant and Aboriginal communities in Canada. Social Science and Medicine. 2005; **61**(5):931-942

[105] De Bruyn T. HIV/AIDS and 1. Discrimination, 2. Stigma and Discrimination: Definitions and Concepts. Ottawa: Canadian HIV/AIDS Legal Network and the Canadian AIDS Society; 1999

[106] Van Rie A, Sengupta S, Pungrassami P, Balthip Q, Choonuan S, Kasetjaroen Y, et al. Measuring stigma associated with tuberculosis and HIV/ AIDS in southern Thailand: Exploratory and confirmatory factor analyses of two new scales. Tropical Medicine and International Health. 2008;**13**(1):21-30. DOI: 10.1111/j.1365-3156.2007.01971.x

[107] Oliva J, Fernández-Bolaños A, Hidalgo A. Health-related quality of life in diabetic people with different vascular risk. BMC Public Health. 2012; **12**:812

[108] Ahari SS, Arshi S, Iranparvar M, Amani F, Siahpush H. The effects of type II diabetes on quality of life. Journal of Ardabil University of Medical Sciences. 2008;**8**(4):394-402

[109] Ahmadi A, Hasanzadeh J, Mediseh MR, Lashkari L. Factors affecting quality of life in patients with type 2 diabetes in Chaharmahal & Bakhtiari province. Journal of North Khorasan University of Medical Sciences. 2011;**3**(1):7-13

[110] Borzou R, Salavati M, Safari M, Hadadinejad SH, Zandieh M, Torkman B. Quality of life in type II diabetic patients referred to Sina Hospital, Hamadan. Zahedan Journal of Research in Medical Sciences. 2010; **13**(4):43-46

[111] Goldney RD, Phillips PJ, Fisher LJ, Wilson DH. Diabetes, depression, and quality of life—A population study. Diabetes Care. 2004;**27**:1066-1070

[112] Larsson D, Lager I, Nilsson PM. Socio-economic characteristics and quality of life in diabetes mellitus/relation to metabolic control. Scandinavian Journal of Public Health. 1999;**27**:101-105

[113] WHO. The World Health Report—Mental Health: New Understanding, New Hope. Geneva, Switzerland: WHO; 2001. Available from: http://www.who.int/whr/2001/en/

[114] Demi A, Bakeman R, Richard S, Linda M, Brenda S. Suicidal thoughts of women with HIV infection: Effect of stressors and moderating effects of family cohesion. Journal of Family Psychology. 1998;**12**:344-353

[115] Caldwell CB, Gottesman II. Schizophrenics kill themselves too: A review of risk factors for suicide. Schizophrenia Bulletin. 1990;**16**:571-589

[116] Dutta R, Boydell J, Kennedy N, Van Os J, Fearon P, et al. Suicide and other causes of mortality in bipolar disorder: A longitudinal study. Psychological Medicine. 2007;**37**:839-847

[117] McGirr A, Turecki G. What is specific to suicide in schizophrenia disorder? Demographic, clinical and behavioural dimensions. Schizophrenia Research. 2008;**98**:217-224

[118] Whitlock G, Lewington S, Sherliker P, Clarke R, Emberson J, et al. Body-mass index and cause-specific mortality in 900 000 adults: Collaborative analyses of 57 prospective studies. Lancet. 2009;**373**:1083-1096

[119] Baxter DN. The mortality experience of individuals on the Salford Psychiatric Case Register. I. All-cause mortality. The British Journal of Psychiatry. 1996;**168**:772-779

[120] Fagiolini A, Goracci A. The effects of undertreated chronic medical illnesses in patients with severe mental disorders. The Journal of Clinical Psychiatry. 2009;**70**(Suppl):322-329

[121] De Hert M, Dekker JM, Wood D, Kahl KG, Holt RI, et al. Cardiovascular disease and diabetes in people with severe mental illness position statement from the European Psychiatric Association (EPA), supported by the European Association for the Study of Diabetes (EASD) and the European Society of Cardiology (ESC). European Psychiatry. 2009;**24**:412-424

[122] Brown S, Birtwistle J, Roe L, Thompson C. The unhealthy lifestyle of people with schizophrenia. Psychological Medicine. 1999;**29**:697-701

Chapter 5

Crude Birth Rate and Crude Mortality Rate in India: A Case of Application of Regression in Healthcare

Prisilla Jayanthi and Muralikrishna Iyyanki

Abstract

India's demographic transition in 1950 has led to decline of high birth and mortality rate to 6.5%, a drop of 0.6% in 2016–2017, as per the economic survey (2017–2018). The crude birth rate and crude mortality rate decreases with the occupation. In this study, the statistical parameter, confidence interval indicates the true range of the mean of the crude birth rate and crude mortality rate computed from the observed data in the study. Location and precision of a measure are made available with the confidence interval. In the study, the results for crude birth rate in 1984 were highest, 95% CI = 32.08–39.22, and in 2011, were lowest, 95% CI = 20.68–25.24, and the results for crude mortality rate in 1984 were highest, 95% CI = 32.08–39.22, and in 2011 were lowest, 95% CI = 20.68–25.24. A small standard error implies that the sample mean is a more accurate reflection of the actual population mean. The smallest standard error of crude birth rate is 1.08, and the smallest standard error of crude mortality rate is 0.50.

Keywords: crude birth rate, crude mortality rate, confidence interval, health

1. Introduction

In statistics, the two statistical values used to measure the growth or decline of a population are crude birth rate (CBR) and crude death rate (CBR). The growth or decline of a population is based on literacy and economic growth of any country. In India, several states have the need for developing itself in measure of health or in capital wise. Agriculture happens to be largest occupation in India where the children at very young age get to work in agriculture leading to large birth rates number in the rural states families. Under the age of 5 years, the crude mortality rate occurs due to preterm birth (18%), pneumonia (16%), interpartum (12%), neonatal sepsis (7%), diarrhea (8%), malaria (5%) and malnutrition. In most of the undeveloped countries, the malnutrition is the major cause of child mortality. The present urban states have better health, hygiene and sanitation facilities in the area lead to decline in mortality rate. Literacy and the economical understanding have blended to lower the birth rate in both rural and urban.

2. Previous study

Golley and Toyer [1] suggested that China and India's demographic transitions timings and the implications of fertility developments were discovered using a global economic model and measures of dependency include the working over-aged and working age. China's labor force to begin to diminish, whereas India will increase fertility rate faster than its present population. The population plays a significant role in defining the relative magnitudes of labor force growth to total population growth and the change in dependency ratios, with a significant impact on per capita income growth. India, the world's most populous country by 2030, and its population policy continue to be directed toward promoting fertility decline. The lower fertility reduces GDP and increases per capita income in both countries, India gains more per capita income than China per unit change in fertility, resulting in India's higher youth dependency [1].

Roy and Jones [2] developed a technique for the prediction of health indicators for all the districts of India and examine the correlations between health and development. The two fundamental indicators of this research are the levels of electrification and district domestic product (DDP). The data with health metrics and the information from two night time satellite images were used to propose the models. The predicted the health indicators with less than 7–10% errors were successfully. The health metrics, like crude birth rate and maternal mortality rate were mapped for the whole country at the area level. These metrics showed very strong correlation with development indicators. In a socio-economic study, using Visible Infrared Imaging Radiometer Suite (VIIRS) satellite imagery, the observation showed a higher DDP and level of electrification for better health conditions [2].

Maitra and Pal [3] emphasized that the estimates of birth spacing on child mortality are different when fertility selection are not considered. A comparison study of the fertility behavior of households in the Indian and the Pakistani Punjab highlighted the differential nature of institutions on demographic transition in these neighboring regions. The study involved reported birth interval and not inter-conception interval, which implies that there were some measurement errors associated with this particular variable. The miscarriages, stillbirths and also premature births were not measured for measurement. The study identifies the bivariate probit model that estimates mortality after correcting for the self-selection in fertility decisions [3].

3. Discussions, empirical analysis and results

The true population value can be calculated using a confidence interval (CI) in statistics, an interval estimate, computed from the observed data [4]. The interval contains the true value of an unknown population parameter with $(1 - \alpha)\%$ confidence which quantifies the level of confidence that the parameter lies in the interval. And the confidence level represents frequency and that are constructed from an infinite number of independent sample statistics, the proportion of those intervals that contain the true value of the parameter will be equal to the confidence level.

The study was carried out using Stata 12.1 IC software to calculate the confidence interval to understand the crude birth rate and crude mortality rate in India in six different states namely Andhra Pradesh, Assam, Madhya Pradesh, Uttar Pradesh, Dadra & Nagar Haveli the data was taken from www.data.gov.in [4] and was collected from the year 1981 to 2011.

Year	Population	Crude birth rate	Crude mortality rate	Percentage decadal variation	Annual exponential growth rate (%)
1971	548.2	41.2	19	24.8	2.2
1981	683.3	37.2	15	24.66	2.22
1991	846.4	32.5	11.4	23.87	2.16
2001	1028.7	24.8	8.9	21.54	1.97
2011	1210.9	21.8	7.1	17.7	1.63

Table 1.
Crude birth and mortality rate in India from the year 1971 to 2011.

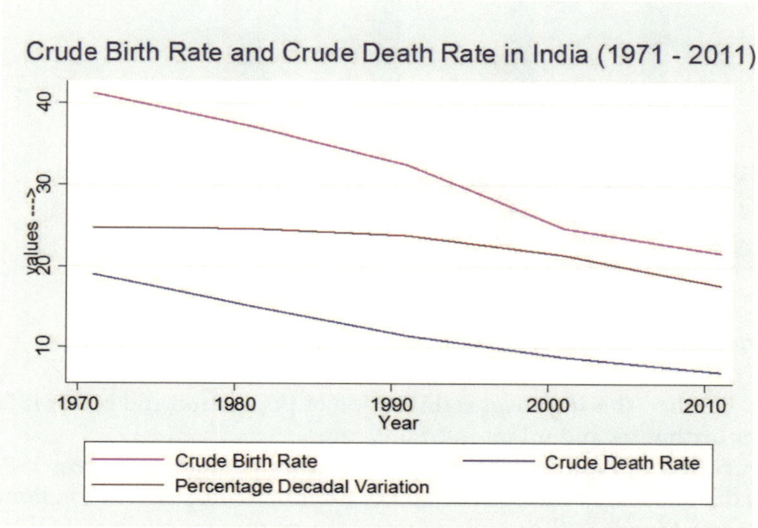

Figure 1.
Crude birth rate and crude mortality rate in India.

Table 1 shows the decrease in crude birth rate and crude mortality rate from 1971 to 2011 with the industrial rise. The graph in **Figure 1** shows the decline in crude birth rate (21.8) and crude mortality rate (7.1) in the year 2011.

From **Table 1**, the population mean = 865.17, SE = 2.37 with 95% CI = 860.52–869.83 and crude birth rate mean = 31.44, SE = 0.07 and 95% CI = 31.30–31.59. The crude mortality rate mean = 12.24, SE = 12.16 showed 95% CI = 0.04–12.33. The results of percentage decadal variation 95% CI = 22.44–22.54, mean = 22.49 and SE = 0.02. The results for annual exponential growth rate were 95% CI = 2.03–2.03, mean = 2.03 and SE = 0.01.

Figure 2 displays the crude birth and death rates for six individual states in India for the period (1971–2011). The peak crude birth rate can be seen in Dadra and Nagar Haveli with 27.7 births per thousand in urban areas. Interestingly, this State also had the lowest death rates in both urban and rural areas (**Figure 2**). The lowest crude birth rates were found in Pondicherry, where both rates were urban is 14 and rural is 13.6. There was greater variability in birth rates in both urban and rural locales than in death rates in either type of setting. The female literacy in any state shows that the crude birth rate and mortality rate is minimum (**Table 2**). Based on the census [5], the educated females take better reproductive and healthcare

Variable	Mean	Std. err	95% CI
Crude birth rate rural	21.31	2.25	15.50–27.12
Crude birth rate urban	19.13	2.17	13.5–24.71
Crude death rate rural	7.08	0.43	5.97–8.18
Crude death rate urban	5.18	0.49	3.90–6.46

Table 2.
CI of crude birth rate and mortality rate in rural and urban (1971–2011).

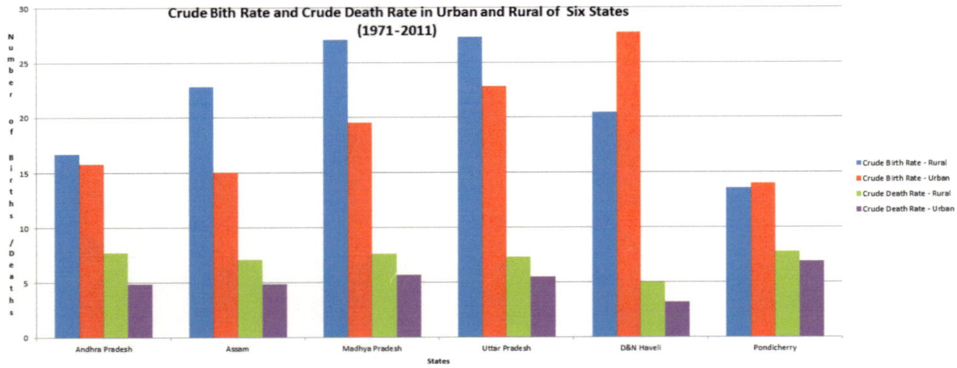

Figure 2.
The graph representing crude birth rate and crude mortality rate in urban and rural.

decisions. Further, this improves stabilization of population and better infant care with lower birth rates and infant mortality rates.

In **Figure 3**, the graph shows the highest crude birth rate in the year 1984 with 45.9 in Dadra and Nagar Haveli state and minimum in the year 2007 in Pondicherry. Similarly from **Figure 4**, the graph indicates the maximum crude mortality rate to be 17.8 in the year 1984 in Uttar Pradesh, and the minimum crude mortality rate was found thrice as 7, twice in the year 2001 and 2009 in Pondicherry and one time in Andhra Pradesh (2004).

In **Figures 5** and **6**, the graph represents the crude birth rate and crude mortality rate in rural and urban from the year 1981 to 2011. The standard error of the

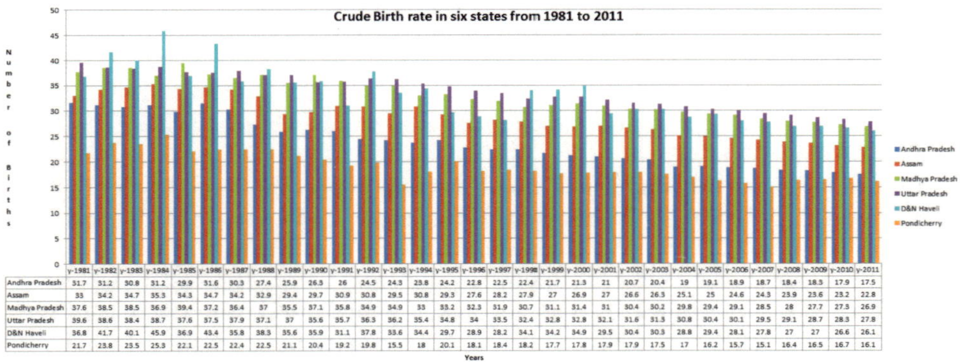

	y-1981	y-1982	y-1983	y-1984	y-1985	y-1986	y-1987	y-1988	y-1989	y-1990	y-1991	y-1992	y-1993	y-1994	y-1995
Andhra Pradesh	31.7	31.2	30.8	31.2	29.9	31.6	30.3	27.4	25.9	26.3	26	24.5	24.3	23.8	24.2
Assam	33	34.2	34.7	35.3	34.3	34.7	34.2	32.9	29.4	29.7	30.9	30.8	29.5	30.8	29.3
Madhya Pradesh	37.6	38.5	38.5	36.9	39.4	37.2	36.4	37	35.5	37.1	35.8	34.9	34.9	33	33.2
Uttar Pradesh	39.6	38.6	38.4	38.7	37.6	37.5	37.9	37.1	37	35.6	35.7	36.3	36.2	35.4	34.8
D&N Haveli	36.8	41.7	40.1	45.9	36.9	43.4	35.8	38.3	35.6	35.9	31.1	37.8	33.6	34.4	29.7
Pondicherry	21.7	23.8	23.5	25.3	22.1	22.5	22.4	22.5	21.1	20.4	19.2	19.8	15.5	18	20.1

	y-1996	y-1997	y-1998	y-1999	y-2000	y-2001	y-2002	y-2003	y-2004	y-2005	y-2006	y-2007	y-2008	y-2009	y-2010	y-2011
Andhra Pradesh	22.8	22.5	22.4	21.7	21.3	21	20.7	20.4	19	19.1	18.9	18.7	18.4	18.3	17.9	17.5
Assam	27.6	28.2	27.9	27	26.9	27	26.6	26.3	25.1	25	24.6	24.3	23.9	23.6	23.2	22.8
Madhya Pradesh	32.3	31.9	30.7	31.1	31.4	31	30.4	30.2	29.8	29.4	29.1	28.5	28	27.7	27.3	26.9
Uttar Pradesh	34	33.5	32.4	32.8	32.8	32.1	31.6	31.3	30.8	30.4	30.1	29.5	29.1	28.7	28.3	27.8
D&N Haveli	28.9	28.2	34.1	34.2	34.9	29.5	30.4	30.3	28.8	29.4	28.1	27.8	27	27	26.6	26.1
Pondicherry	18.1	18.4	18.2	17.7	17.8	17.9	17.9	17.5	17	16.2	15.7	15.1	16.4	16.5	16.7	16.1

Years

Figure 3.
The graph representing six states' crude birth rate.

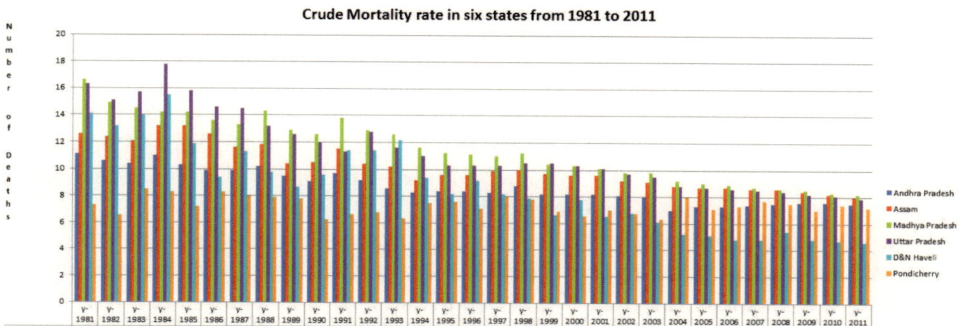

	Y-1981	Y-1982	Y-1983	Y-1984	Y-1985	Y-1986	Y-1987	Y-1988	Y-1989	Y-1990	Y-1991	Y-1992	Y-1993	Y-1994	Y-1995
Andhra Pradesh	11.1	10.6	10.4	11	10.3	9.9	9.9	10.2	9.5	9.1	9.7	9.2	8.6	8.3	8.4
Assam	12.6	12.4	12.1	13.2	13.2	12.6	11.6	11.8	10.4	10.5	11.5	10.4	10.2	9.2	9.6
Madhya Pradesh	16.6	14.9	14.5	14.2	14.2	13.6	13.3	14.3	12.9	12.6	13.8	12.9	12.6	11.6	11.2
Uttar Pradesh	16.3	15.1	15.7	17.8	15.8	14.6	14.5	13.2	12.6	12	11.3	12.8	11.6	11	10.3
D&N Haveli	14.1	13.2	14	15.5	11.9	9.4	11.3	9.8	8.7	9.6	11.4	11.4	12.2	9.4	8.2
Pondicherry	7.3	6.5	8.5	8.3	7.2	8.3	8	7.9	7.8	6.2	6.6	6.8	6.3	7.5	7.6

	Y-1996	Y-1997	Y-1998	Y-1999	Y-2000	Y-2001	Y-2002	Y-2003	Y-2004	Y-2005	Y-2006	Y-2007	Y-2008	Y-2009	Y-2010	Y-2011
Andhra Pradesh	8.4	8.3	8.8	8.2	8.2	8.2	8.1	8	7	7.3	7.3	7.4	7.5	7.6	7.6	7.5
Assam	9.6	9.9	10	9.7	9.6	9.6	9.2	9.1	8.8	8.7	8.7	8.6	8.6	8.4	8.2	8
Madhya Pradesh	11.1	11	11.2	10.4	10.3	10.1	9.8	9.8	9.2	9	8.9	8.7	8.6	8.5	8.3	8.2
Uttar Pradesh	10.3	10.3	10.5	10.5	10.3	10.1	9.7	9.5	8.8	8.7	8.6	8.5	8.4	8.2	8.1	7.9
D&N Haveli	9.2	8.2	7.9	6.6	7.8	6.5	6.8	6.1	5.2	5.1	4.8	4.8	5.4	4.8	4.7	4.6
Pondicherry	7.1	8	7.8	6.9	6.5	7	6.7	6.3	8	7.1	7.3	7.7	7.5	7	7.4	7.2

Years

Figure 4.
The graph representing crude mortality rate in six states (1981–2011).

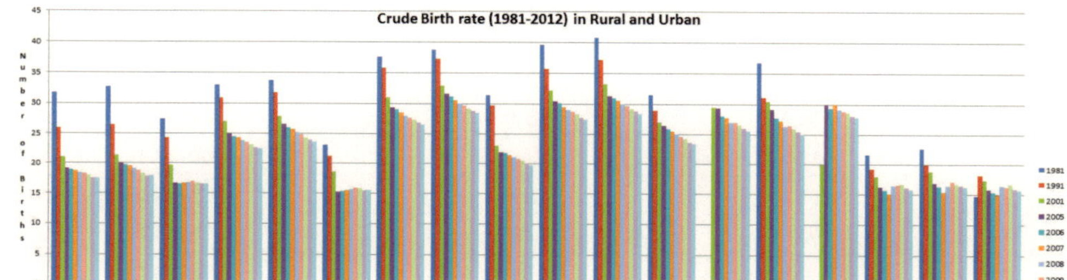

	Andhra Pradesh-Total	Andhra Pradesh-Rural	Andhra Pradesh-Urban	Assam -Total	Assam -Rural	Assam -Urban	Madhya Pradesh -Total	Madhya Pradesh -Rural	Madhya Pradesh -Urban
1981	31.7	32.7	27.5	33	33.8	23.2	37.6	38.8	31.4
1991	26	26.5	24.4	30.9	31.7	21.3	35.8	37.3	29.7
2001	21	21.4	19.7	27	27.9	18.6	31	32.9	23.1
2005	19.1	20.1	16.7	25	26.6	15.3	29.4	31.6	22
2006	18.9	19.8	16.5	24.6	26.1	15.4	29.1	31.2	21.9
2007	18.7	19.5	16.7	24.3	25.7	15.5	28.5	30.5	21.5
2008	18.4	19.1	16.8	23.9	25.3	15.7	28	30	21.1
2009	18.3	18.8	17	23.6	24.9	15.9	27.7	29.7	20.8
2010	17.9	18.3	16.7	23.2	24.4	15.8	27.3	29.2	20.5
2011	17.5	17.8	16.6	22.8	24	15.5	26.9	28.8	20.1
2012	17.5	17.9	16.6	22.5	23.7	15.6	26.6	28.5	19.8

	Uttar Pradesh -Total	Uttar Pradesh -Rural	Uttar Pradesh -Urban	Dadra & Nagar Haveli -Total	Dadra & Nagar Haveli-Urban	Dadra & Nagar Haveli-Rural	Pondicherry -Total	Pondicherry-Urban	Pondicherry-Rural
1981	39.6	40.8	31.5	0	36.8	0	21.7	22.8	14.7
1991	35.7	37.2	29	0	31.1	0	19.2	20	18.1
2001	32.1	33.2	27	29.5	30.4	20	17.9	18.8	17.3
2005	30.4	31.3	26.5	29.4	29.2	30	16.2	16.9	15.8
2006	30.1	31	26	28.1	27.8	29.4	15.7	16.3	15.4
2007	29.5	30.5	25.5	27.8	27.2	30	15.1	15.4	14.9
2008	29.1	30	25.1	27	26.4	29.2	16.4	16.4	16.4
2009	28.7	29.7	24.7	27	26.5	28.8	16.5	17.1	16.2
2010	28.3	29.2	24.2	26.6	26	28.6	16.7	16.7	16.7
2011	27.8	28.8	23.7	26.1	25.5	28.1	16.1	16.4	15.9
2012	27.4	28.4	23.5	25.6	25.1	27.9	15.8	16.2	15.7

States

Figure 5.
The graph representing crude birth rate in rural and urban in six states (1981–2011).

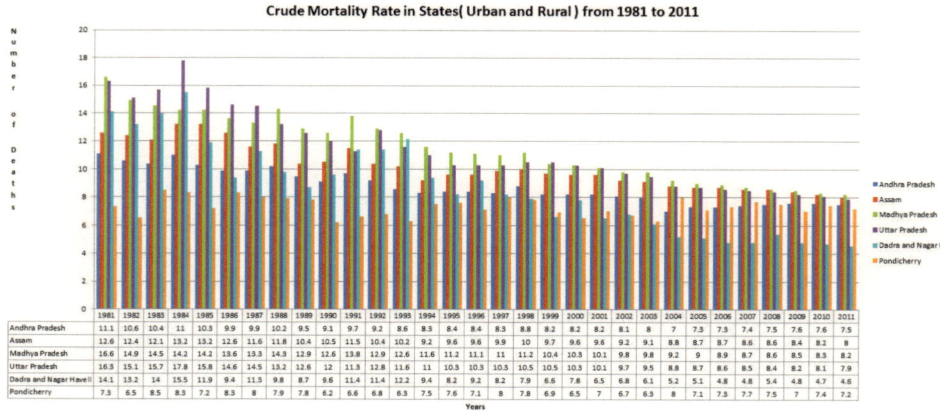

	1981	1982	1983	1984	1985	1986	1987	1988	1989	1990	1991	1992	1993	1994	1995
Andhra Pradesh	11.1	10.6	10.4	11	10.3	9.9	9.9	10.2	9.5	9.1	9.7	9.2	8.6	8.3	8.4
Assam	12.6	12.4	12.1	13.2	13.2	12.6	11.6	11.8	10.4	10.5	11.5	10.4	10.2	9.2	9.6
Madhya Pradesh	16.6	14.9	14.5	14.2	14.2	13.6	13.3	14.3	12.9	12.6	13.8	12.9	12.6	11.6	11.2
Uttar Pradesh	16.3	15.1	15.7	17.8	15.8	14.6	14.5	13.2	12.6	12	11.3	12.8	11.6	11	10.3
Dadra and Nagar Haveli	14.1	13.2	14	15.5	11.9	9.4	11.3	9.8	8.7	9.6	11.4	11.4	12.2	9.4	8.2
Pondicherry	7.3	6.5	8.5	8.3	7.2	8.3	8	7.9	7.8	6.2	6.6	6.8	6.3	7.5	7.6

	1996	1997	1998	1999	2000	2001	2002	2003	2004	2005	2006	2007	2008	2009	2010	2011
Andhra Pradesh	8.4	8.3	8.8	8.2	8.2	8.2	8.1	8	7	7.3	7.3	7.4	7.5	7.6	7.6	7.5
Assam	9.6	9.9	10	9.7	9.6	9.6	9.2	9.1	8.8	8.7	8.7	8.6	8.6	8.4	8.2	8
Madhya Pradesh	11.1	11	11.2	10.4	10.3	10.1	9.8	9.8	9.2	9	8.9	8.7	8.6	8.5	8.3	8.2
Uttar Pradesh	10.3	10.3	10.5	10.5	10.3	10.1	9.7	9.5	8.8	8.7	8.6	8.5	8.4	8.2	8.1	7.9
Dadra and Nagar Haveli	9.2	8.2	7.9	6.6	7.8	6.5	6.8	6.1	5.2	5.1	4.8	4.8	5.4	4.8	4.7	4.6
Pondicherry	7.1	8	7.8	6.9	6.5	7	6.7	6.3	8	7.1	7.3	7.7	7.5	7	7.4	7.2

Years

Figure 6.
The graph representing crude mortality rate in urban and rural in six states (1981–2011).

mean decreases as the size of observations increases. A small standard error implies that the sample mean is more accurate replication of the actual population mean. The crude birth rate with the least SE = 1.08 and in crude mortality rate, the least SE = 0.49. **Table 3** refers to highest crude birth rate 95% CI = 32.07–39.21 (1984) and least crude birth rate 95% CI = 20.67–25.23 (2011). Likewise, **Table 4** indicates highest crude mortality rate of 95% CI = 9.813–16.85 in 1984 and least crude mortality rate of 95% CI = 5.82–8.63 in the year 2011.

The crude mortality rate in children under the age of 5 years occurs due to preterm birth, pneumonia, neonatal sepsis, diarrhea, malaria and malnutrition. In undeveloped countries, malnutrition is the primary cause of child mortality.

In India, the highest total crude birth rate was in the year 1981 with 35.6 in rural and minimum in the year 2012 with 17.4 (urban).

Year	Mean	Standard error	95% conf. interval
1981	32.53	1.59	29.19–35.86
1982	34.32	1.59	30.99–37.65
1983	33.85	1.54	30.63–37.06
1984	35.65	1.71	32.07–39.21
1985	32.58	1.54	29.35–35.80
1986	34.03	1.77	30.32–37.72
1987	32.04	1.41	29.10–34.98
1988	32.34	1.49	29.21–35.45
1989	30.66	1.48	27.55–33.75
1990	30.54	1.55	27.30–33.77
1991	28.98	1.49	25.87–32.09
1992	30.65	1.67	27.15–34.15
1993	28.27	1.90	24.30–32.24
1994	28.86	1.63	25.45–32.25
1995	28.12	1.26	25.48–30.76
1996	26.85	1.37	23.99–29.72
1997	26.66	1.30	23.94–29.38
1998	27.6	1.45	24.56–30.63
1999	27.49	1.53	24.29–30.69
2000	27.70	1.56	24.43–30.97
2001	26.25	1.29	23.55–28.95
2002	26.23	1.29	23.53–28.92
2003	25.96	1.31	23.23–28.70
2004	25.13	1.28	22.44–27.81
2005	24.90	1.35	22.08–27.73
2006	24.30	1.33	21.52–27.09
2007	23.83	1.34	21.02–26.62
2008	23.81	1.17	21.37–26.24
2009	23.68	1.14	21.31–26.05
2010	23.45	1.08	21.19–25.71
2011	22.95	1.09	20.67–25.23

Table 3.
Mean, standard error and 95% confidence interval for crude birth rate in India, by year (1981–2011).

Confidence interval for crude mortality rate (1981–2011)			
Year	**Mean**	**SE**	**95% conf. interval**
1981	13	1.42	9.32–16.67
1982	12.12	1.31	8.74–15.49
1983	12.53	1.11	9.67–15.39
1984	13.33	1.36	9.81–16.85
1985	12.1	1.24	8.89–15.30
1986	11.4	1.03	8.72–14.07
1987	11.43	0.94	8.99–13.87
1988	11.2	0.96	8.72–13.67
1989	10.31	0.84	8.14–12.49
1990	10	0.94	7.58–12.41
1991	10.71	0.98	8.19–13.24
1992	10.58	0.95	8.13–13.03
1993	10.25	0.99	7.70–12.79
1994	9.5	0.64	7.86–11.13
1995	9.21	0.56	7.76–10.66
1996	9.28	0.57	7.80–10.76
1997	9.28	0.52	7.94–10.62
1998	9.36	0.57	7.88–10.84
1999	8.71	0.70	6.89–10.53
2000	8.78	0.62	7.17–10.39
2001	8.58	0.64	6.91–10.25
2002	8.38	0.57	6.91–9.85
2003	8.13	0.66	6.43–9.83
2004	7.83	0.61	6.24–9.41
2005	7.65	0.60	6.09–9.20
2006	7.6	0.63	5.97–9.22
2007	7.61	0.60	6.06–9.16
2008	7.66	0.49	6.38–8.94
2009	7.41	0.57	5.94–8.88
2010	7.38	0.55	5.95–8.81
2011	7.23	0.54	5.82–8.63

Table 4.
Confidence interval, mean and standard error for crude mortality rate in India.

4. Demographic transition in India

In India, when the country became a republic in the 1950s, it experienced a demographic transition from high birth rate and mortality rate to lower birth rate and lower mortality rate. The country revolutionized from pre-industrial to an industrialized economic system. This transition brought several changes in a lower fertility rate where smaller, independent families emerged and more resources saved and invested in capital and education. More investment contributes to economic growth. This transition led to rise in socio-economic relations, health impacts and India's economic level.

5. Conclusion

From the above discussion the highest crude birth rate was 45.9 (1984) and minimum crude birth rate was 15.1 (2007) and the maximum crude mortality rate to be 17.8 (1984) minimum crude mortality rate was 7 (2004). The decline in crude birth rate (21.8) and crude mortality rate (7.1) in the year 2011 was found. Industrialization brought a drastic change in the economic growth of the people as they started to exchange ideas and get more involved in the change and development. This demographic transition in India led to enhanced education levels for women in India (2011) and hence decreasing crude birth rate and mortality rate. For the year 2011, the percentage distribution of second and higher order live births by interval between current and previous live birth has been shown in few states of India. The spacing between children in the rural and urban areas implies that about half of the birth should have spacing 36 months and above. Most of rural and urban areas now have 70% of births which have birth interval of 24 and more months. The study shows the confidential interval for the highest crude birth rate 95% CI = 32.07–39.21 (1984) and least crude birth rate 95% CI = 20.67–25.23 (2011). Likewise, the highest crude mortality rate of 95% CI = 9.813–16.85 in 1984 and least crude mortality rate of 95% CI = 5.82–8.63 in the year 2011 can be noted. These are the key factors for the crude birth and crude mortality rate to decline from 1981 to 2011. The confidence interval and lower SE helps to get the accurate mean of the population in a particular region with a range.

Author details

Prisilla Jayanthi[1*] and Muralikrishna Iyyanki[2]

1 KG Reddy College of Engineering and Technology, Moinabad, India

2 R&D, JNTUH, Hyderabad, India

*Address all correspondence to: prisillaj28@gmail.com

References

[1] Golley J, Toyer R. Contrasting giants: Demographic change and economic performance in China and India. Procedia-Social and Behavioral Sciences. 2011. DOI: 10.2139/ssrn.1829652

[2] Roychowdhury K, Jones S. Nexus of health and development: Modelling crude birth rate and maternal mortality ratio using nighttime satellite images. ISPRS International Journal of Geo-Information. 2014;**3**:693-712. DOI: 10.3390/ijgi3020693. ISSN 2220-9964

[3] Maitra P, Pal S. Birth spacing, fertility selection and child survival: Analysis using a correlated hazard model. Journal of Health Economics. 2008;**27**:690-705

[4] ci–Stata. Using Stata for Confidence Intervals. Available from: https://www.stata.com/manuals13/rci.pdf

[5] Saurabh S, Sarkar S, Pandey DK. Female literacy rate is a better predictor of birth rate and infant mortality rate in India. Journal of Family Medicine and Primary Care. 2013;**2**(4):349-353. DOI: 10.4103/2249-4863.123889